HISTORY OF SCIENCE AS EXPLANATION

HISTORY OF SCIENCE AS EXPLANATION

by
Maurice A. Finocchiaro

WAYNE STATE UNIVERSITY PRESS □ DETROIT, 1973

*Published simultaneously in Canada
by the Copp Clark Publishing Company
517 Wellington Street, West Toronto 2B, Canada.*

Library of Congress Cataloging in Publication Data
Finocchiaro, Maurice A   1942–
  History of Science as explanation.
  Bibliography: p. 276–79
  1. Science—History. 2. Science—Philosophy.
I. Title.
Q125.F49 1972   509   72–3582
ISBN 0–8143–1480–5

ACKNOWLEDGMENTS

*Grateful acknowledgment is extended to the publishers for permission to quote
from the following works:*

  George Basalla, The Rise of Modern Science, *Lexington, Mass.: D. C. Heath and
Company, 1968.*

  G. N. Clark, Science and Social Welfare in the Age of Newton, 2d ed., Oxford,
Clarendon Press, 1949.

  Benedetto Croce, History as the Story of Liberty, *translated by Sylvia Sprigge,
London: George Allen and Unwin, 1941;* Theory and History of Historiography,
London: George G. Harrap & Co., 1921.

  Alexandre Koyré, Etudes Galiléennes (1939), *Paris: Hermann, 1971;* Metaphysics and Measurement, *London: Chapman & Hall, 1968;* Metaphysics and
Measurement, *Cambridge: Harvard University Press, 1968.*

To Benedetto Croce, Paul Feyerabend,
Karl Popper, and Michael Scriven, none
of whom might like the others' company
but each of whom I have tried in my
own way to emulate.

contents

Contents

chapter **1**

Explanation and the Critical Philosophy
of the History of Science

WHAT DISTINGUISHES PHILOSOPHY from other disciplines and activities is its concern with *critical understanding*. All others are either not essentially concerned with understanding but with such things as generalization, abstraction, prediction, and systematization, or they are concerned with understanding which is non-critical or even uncritical. The critical philosophy of the history of science aims primarily at the critical understanding of the development of science, and ultimately at the critico-historical understanding of science itself. Subordinately, it aims at the critical understanding of the historiography of science[1] since the very least that the critical understanding of the development of science and the critico-historical understanding of science require is an awareness of the methods for acquiring understanding and of the

9

very concept of understanding. The concepts and methods needed in the study of the history of science, then, themselves need some study.

But there is a less philosophical and more pragmatic reason for such needed attention. In fact, if there is anything which is universally agreed upon about the study of the history of science, it is the socio-institutional fact that in recent times it has become institutionalized into a profession. Evidence for this is provided by the increasing number of university courses dealing exclusively with the history of science and of university teachers giving—primarily or exclusively—courses on the subject, and by the emergence of university departments of History of Science and of Ph.D. programs in history of science.

The most immediate and inevitable result of such developments will be various higher level studies which will regard them as their raison d'être. For example, the sociologist will investigate the sociological structure of the new group and the social functions and dysfunctions of various aspects of that structure.

A second kind of investigation that will emerge is the study of the history or, if you will, of the prehistory of the discipline. Such a historian of the historiography of science will investigate its origins and trace its development in time. The historian's interests will partly overlap with those of the sociologist, but no problem need arise from this dual attention.

A third kind of investigation will be the attempt to describe, understand, and evaluate the intellectual activities of the newly formed group. This type of inquiry may be called philosophical or methodological. Thus what may be called the methodology of the historiography of science or the critical philosophy of the history of science will evolve.

Joseph Agassi's *Towards an Historiography of Science*[2] is a pioneer work on the subject. It will be a long time, and if Agassi's influence prevails, an infinitely long time indeed, before the methodology of history of science itself becomes professionalized. Be that as it may, let us reflect upon the relations between it and the historiography of science, remembering that a raison d'être of this methodology *is* the historiography of science.

Hence, in evaluating the philosopher Agassi's pioneer contribution, my reaction to historians' criticism of it will be neither laughter nor dismissal, "laughter" being one historian's implied reaction to criticism of the historiography of science by scientists,[3] and "dismissal" what another historian claims would be a justified reaction to Agassi's work on the part of historians of science.[4] Instead, my reaction will be *rational* criticism.

Consider the arguments that generally speaking a history of science cannot be competently judged by a scientist because of his prejudices, caused by the ideology of the scientific profession; or because those histories are now being written in such a manner that only the professional scholar can understand them. I will not generalize these arguments so as to make them applicable at the level of the relation between the historiography of science and its methodology. This attitude, consequently, will put the historian at an advantage, yet at the same time it will constitute an oblique criticism of him, since I will reject those arguments when applied at the level of the relation between science and history of science. For though it *may* be necessary for history of science to extricate itself from general history, for it to isolate itself from or ignore the reactions of scientists is as disastrous and as doomed to failure as the attempts of general historians to isolate themselves from the mass of intelligent and educated laymen. In other words, the scientific community plays the same role for historians of science that the human community plays for general historians.

For at least two reasons historians of science then should have no reason to resent the existence of "presumptuous self-appointed critics," as one historian of science characterizes Agassi.[5] The first is that the emergence of such self-appointed critics is a direct consequence of the institutionalization of the discipline, for which *they* are responsible. But, more importantly, such critics will not be critics simply in the sense of "destructive critics," but instead will be concerned with critical understanding. They will no less frequently emphasize the good as well as the bad aspects. Moreover, they will often criticize each other. Conversely, if one methodologist makes criticism which historians of science by and large find to be unfair, and which they cannot coherently and rationally answer, another methodologist will often be in a better position to do just that—and without compromising his critical aims. I believe the last point applies to the investigations which I shall undertake, since they are in part a defense of history of science from Agassi's criticism, which, as I shall show, has not really been answered by historians of science themselves.

The need to study the concepts and methods required and used by historians of science is then beyond question. What I intend to investigate is the nature and adequacy of a certain intellectual practice used by historians of science—that in which a historian of science is or can be said to be explaining something, that is, *giving an explanation.*

The philosophical importance of the concept of explanation is of course well known. Some philosophers of science, e.g., Ernest Nagel, have attempted to give, by means of it, a logical structure to scientific knowledge.[6] Other philosophers of science, e.g., Agassi, have attempted, by means of the concept of explanation, to give an interpretation of the history of science according to which scientific theories are explanations of facts, and not, for example, inductive generalizations *from* facts or abstract classifications imposed on facts.[7] Positivist methodologists of history have attempted to make historiography dependent on the so-called generalizing or theoretical sciences by making historical explanation dependent on general laws.[8] And in the philosophy of the social sciences, Karl Popper has used what he takes to be the nature of the concept of explanation to argue against a historicist approach to sociology which would make the latter nothing but theoretical history.[9]

One need not accept these diverse views, but the general concern with the concept of explanation can hardly be missed. The importance of explanation in historiography of science, however, has escaped the attention of both philosophers of science and historians of science, with the possible exception of a few hints by Agassi.[10] One reason for its significance is that, as I shall argue in chapters 2, 3, and 4, history-of-science explanation[11] presents serious difficulties for general philosophical principles and analyses of the concept of explanation. From this point of view history-of-science explanation may be said to be anomalous.

A second reason for its importance is that the attempts on the part of historians of science to put forth explanations of various history of science developments have seldom been successful. In other words, as I shall show in chapters 5, 6, and 7, the practice of history-of-science explanation is unsatisfactory.

Third, the concept of history-of-science explanation can indeed serve to state a connection between history of science and philosophy of science. But the connection is not the history of science analog of the positivist thesis, namely, that history of science is dependent on the philosophy of science in that history-of-science explanation is dependent on philosophy-of-science principles. Rather history-of-science explanation will be found in chapter 10 to be at least as autonomous as ordinary historical explanation really is.[12] And it will be certain forms of the philosophy of science which will be shown in chapter 11 to be dependent on history of science. This dependence can be stated by saying that certain history-of-science explanations are the *sources and grounds* for certain principles or claims of the philosophy of science.

Fourth, explanation is the activity in which philosophies of the history of science are at their weakest. Thus it is likely that, as I argue in chapter 12, Thomas Kuhn's theory of scientific change and certain interpretations of Popper's philosophy are barren from the point of view of the explanation of specific history-of-science developments and useful, or misleading, only in trying to solve the problem—if indeed it is a genuine problem—of the whole evolution of science.

Fifth, explanation is, as I shall argue in chapters 13 and 14, the key bridging concept between history, scholarship, and chronicle; that is, the main difference between history and chronicle is that the former contains and the latter lacks explanations, whereas scholarly works may be either histories or chronicles depending on the presence or absence of explanations. This conclusion is supported by and in turn supports a synthesis of Benedetto Croce's theory of history and Michael Scriven's analysis of explanation.

The investigation that follows has its own internal structure, an *explanatory* one. In the first series of chapters (2–7) I shall state what will be called the problem of explanation in historiography of science. In the second group, chapters 8–12, I shall consider Agassi's ideas on historiography as a possible solution and criticize them as unsuccessful. Finally, in the last set of chapters (13–19) I shall make a series of suggestions intended to solve the problem, at least in part.

PART ONE

History-of-Science Explanation

chapter 2

The Explanation of Discoveries: Predictiveness

IN AN ATTEMPT to investigate the nature and adequacy of history-of-science explanation, one of the first questions to be answered, or at least asked, is, What *is* history-of-science explanation? A nominal definition is that it is the explanation that a historian of science may be expected to give. A second and somewhat overlapping definition asserts it to be the explanation of facts and events in the development and evolution of science. These definitions obviously do not take us very far, but they do allow us to begin. For they do tell us that history-of-science explanation is the explanation of a special set of facts and events, so that the nature of explanation in this special field will be a special case of the nature of explanation in general. Thus, if we want to get some insight into the nature of history-of-science explanation, one approach would be to try to apply some of the general principles about the nature of explanation.

One such alleged principle is the following: all satisfactory explanations of events are such that *if* the explanatory information had been available and taken into account before the occurrence of the event *then* it could have been used or would have been sufficient to inferentially predict the event.[1] For brevity, we may regard *potential predictiveness* as that property of an explanation of an event such that *if* the explanatory information had been available and taken into account before the occurrence of the event *then* it could have been used or would have been sufficient to predict the event inferentially. The above mentioned thesis may then be expressed by saying that all satisfactory explanations of events are potentially predictive, and it will be called the *Potential Predictiveness Principle for Explanation*, the PPP for short. In the statement of the PPP the expression "to predict inferentially" has been used. This expression I am using as synonymous with "to predict with deductively certain or with inductively probable arguments." The contrasts I have in mind are the ones with (1) prediction based on skilled judgment, (2) prediction without reasons or arguments, and (3) guesses.

This principle I wish indeed to apply to history-of-science explanation, in the spirit of those who put it forth, for example, Carl Hempel. Such an application is impossible without carefully elucidating the meaning, the rationale, and the difficulties encountered in the more common applications of the principle.

Clarification of the PPP

Good examples of potentially predictive explanations are found in astronomy; here the explanations of solar and lunar eclipses, after their occurrence, usually consist of the same arguments which allowed one to predict them inferentially prior to their occurrence. Potential predictiveness is also well exhibited by some explanations of everyday life. Think, for example, of the explanation a child might give an irate parent who has asked why there is a stain on the new rug: a milkshake was dropped on it. Finally, we find other instances in history, though history obviously is not concerned with actual predictions. For example, we might explain the wreckage and misery that the city of Lisbon experienced on a certain day in the eighteenth century by saying that there had been a strong earthquake a short time before.

And we might explain the wreckage and misery experienced by the city of Hiroshima on a certain day in 1945 by saying that an atom bomb had been dropped on it the day before. In these cases it is obvious that we would have good reasons to predict the wreckage and misery *if* we had known about the occurrence of the causes; and in the case of Hiroshima the American military men did in fact predict them, confident that they would bring about the cause.

The PPP should be distinguished from the proposition that all satisfactory explanations of events are such that the explanatory information could have been used to predict the event inferentially, which thesis might be called the *predictivist requirement for explanations.* The difference between this requirement and the PPP is, of course, the absence in the former and presence in the latter of the 'if' clause. We might state this difference by saying that, whereas the PPP claims that explanations are hypothetically or conditionally predictive, the predictivist requirement asserts that they are actually predictive. Of course, any actually predictive explanation is potentially predictive, but not every potentially predictive one is actually predictive.

The PPP also should be distinguished from the so-called symmetry thesis for explanations and predictions, namely, the proposition that the logical structure of explanation and of prediction is identical. This symmetry claim is itself difficult to interpret, but, whatever its meaning, it states the identity or at least the structural similarity between two things, whereas the PPP attributes an allegedly essential property to explanation. Moreover, whereas the symmetry thesis speaks of explanation and prediction in general, the PPP refers exclusively to the explanation and (potential) prediction of events. Finally, whereas the symmetry thesis is only descriptive of the structure of explanation and prediction, the PPP is supposed to have an important normative function as shown by its reference to satisfactory explanations.

Finally, the PPP should be distinguished from what Hempel calls the general condition of adequacy for any rationally acceptable explanation.[2] This condition is difficult to interpret. But whatever it claims, it is supposed to be a ground for the PPP, as we shall soon see in my discussion of the justification of the PPP. The condition is difficult to interpret because it is sometimes stated by Hempel as the requirement that any satisfactory explanation of a particular event must offer information showing that the event was to be expected,[3] and when so stated the condition may be synonymous with the PPP. At other times Hempel states the condition as follows: any satisfactory explanation of a particular event must provide information which constitutes good grounds for the belief that the event did in fact occur.[4] And when so stated the condition is obviously different from the PPP. Therefore, since Hempel refers to both of these statements as "the general condition of adequacy," and since he uses this condition to justify the PPP, the first of these two statements will be interpreted as being synonymous with the second, and not with the PPP. That is, when Hempel says that any satisfactory explanation must offer information which shows that the event was to be expected, I shall take him to mean that any satisfactory explanation must offer us information which constitutes good grounds for expecting or believing that the event did occur.[5]

Justification of the PPP

In support of the PPP there are two arguments: an a priori and an a posteriori one. In the a priori argument the PPP is grounded on what Hempel calls a general condition of adequacy for any rationally acceptable explanation of a particular event. In the a posteriori argument the PPP is grounded on the alleged logical structure of scientific explanations.

Hempel's a priori argument[6] can be reconstructed as follows. An explanation is an answer to a 'why' question. An explanation of an event is thus an answer to the question "Why did the event occur?" Therefore, a satisfactory explanation of an event is an adequate answer to the question why the event occurred. But an adequate answer to the question why the event occurred must show why the event occurred. And in order to show why an event occurred we must provide information which constitutes good grounds for believing that the event did in fact occur. Therefore, a satisfactory explanation of an event must provide information which constitutes good grounds for believing that the event did in fact occur. But to provide information which constitutes good grounds for believing that an event did in fact occur is to provide information constituting good grounds for expecting that the event did in fact occur, and to do the latter is to provide information constituting good grounds for expecting the event, and this in turn is to provide information showing that the event was to be expected. A satisfactory explanation of an event must therefore provide information showing that the event was to be expected. But if an explanation of an event provides information showing that the event was to be expected, it provides information which is sufficient to predict the occurrence of the event with certainty or probability—provided that the information was available at a suitable earlier time, that is to say, the explanation is potentially predictive. Therefore all satisfactory explanations of events are potentially predictive.

The a posteriori argument begins with the analysis of admittedly satisfactory explanations, such as scientific explanations. When specific scientific explanations are analyzed, so the argument goes, one finds that they have what may be called an *inferential structure*, with three features. First there are two main elements in the explanations, the thing (event) to be explained and what does the explaining. The statement describing the thing (event) to be explained may be called the explicandum sentence, while the set of statements doing the explaining may be called the explicans. The second feature is that the explicans consists of two kinds of statements, particular statements describing initial conditions and general statements or laws. Third, the general statements may be universal and together with the initial conditions deductively imply the explicandum sentence; or the general statements may be of a statistical kind and together with the initial conditions make it probable (without deductively implying it) that the explicandum is true. But, so the argument continues, if an explicans implies or supports an explicandum, that means that it can be used to infer the explicandum with necessity or with probability. Therefore, if the information contained in the explicans had been available before the occurrence of the explicandum event, then it could have been used or would have been sufficient to infer the occurrence of the event in advance, i.e., to predict it inferentially.

General Difficulties

In criticism of the PPP there are several objections, the first four of which are objections to its truth, based on the alleged existence of counterexamples to the principle, that is, of satisfactory explanations which are not potentially predictive. The supporters of the PPP have replied by arguing that the objections miss their mark because all alleged counterinstances are either unsatisfactory or potentially predictive after all.

Explanations in Evolutionary Biology. First, one might base an objection to the PPP on the nature of explanation in evolutionary biology. For example, one might claim, with Michael Scriven, that "the most

21

important lesson to be learned from evolutionary theory today is a negative one: the theory shows us what scientific explanations need not do. In particular it shows us that one cannot regard explanations as unsatisfactory . . . when they are not such as to enable the event in question to have been predicted." [7] But, as Scriven himself points out, this only means that explanations in evolutionary biology are not actually predictive, i.e., that "we can *explain why* certain animals and plants survived even when we could not have *predicted that* they would." [8] Hence, even assuming the satisfactoriness of such explanations, they do not falsify the PPP. Indeed, as Scriven himself says, those explanations are "hypothetically probabilistically predictive": "Naturally we could have said in advance that *if* a flood had not occurred they would be *likely* to survive; let us call this a hypothetical probability prediction." [9] Perhaps one should conclude that the consequences of the nature of explanation in evolutionary biology are more methodological than logical, though, for that reason, they are all the more important.

The above discussion has assumed that the explanations referred to are satisfactory. But this too has been questioned by the supporters of PPP. In fact, as Hempel argues, in evolutionary biology one must distinguish two things, the *story* of evolution and the *theory* of mutation and natural selection. [10] The story is the hypothetical historical narrative describing the various stages of evolutionary sequence, such as the appearance and subsequent extinction of a certain species. The theory is what tries to account for and explain the various events and processes described in the story. And one must neither confuse the descriptions in the story with the theoretical explanations nor overestimate the extent to which those explanations are satisfactory. It is well known that the theory cannot explain many of the details of the evolutionary sequence, such as the fact that a certain kind of organism came into existence at a certain time.

Explanations in Quantum Theory. A second objection to the PPP is based on the nature of the explanations provided by quantum theory. Here, it is argued, by Norwood Hanson, for example, that the explanation of an individual microphysical event or phenomenon P, such as the disintegration of an atom in a radioactive substance, is such that even if the information contained in it is available before the event, it does not permit the inferential prediction of the event. [11]

Hempel's own answer to this objection [12] is mistaken. He thinks that such quantum theoretical explanations are not counterexamples to the PPP because they are examples of explanations which are also potential predictions in which the event is predicted not with deductive certainty but with high inductive probability. It may indeed be that the best analysis of such quantum theoretical explanations is to interpret them as inductive explanations and not as complete explanations, as Hanson is inclined to on the grounds that it is quantum theory that provides the meaning of the expression "explanation of a single microevent." [13] We may also agree with Hempel in saying that "the laws of radioactive decay permit the prediction of events such as the emission of beta-particles by disintegrating atoms only with probability and not with deductive-nomological definiteness for an individual occurrence." [14] But we cannot conclude from this fact that "the explanans . . . can show . . . that the occurrence of P was highly probable," [15] that is, that the quantum theoretical explanation of a single microevent P is a potential inductive prediction of P. In fact, we have neither a genuine prediction, nor a prediction of P.

We do not have a real prediction for the following reason. True, we may know that the probability is one-half that a particular atomic disintegration should occur in a time interval corresponding to the half-life of the element involved. But this does not mean that we can inferentially predict that P will occur in the time interval, for it is no more correct to say that we can predict with probability that the disintegration will occur in the time interval, than it is to say that we can predict with probability that it will *not* occur, since the probability involved is one-half. In other words, to argue that an event will occur with slight probability is not to predict its occurrence inferentially. To equate the two would be a misuse of the notion of prediction.

Moreover, even if Hempel were right in saying that one can inferentially predict that a particular disintegration should occur in the half-life time interval, to predict this would be to predict the wrong thing, namely the event described by the sentence "Atom A disintegrates in time interval between t and $t'' > t$"; whereas microevent P in the quantum theoretical explanation refers to the disintegration of a particular atom at a particular time t' between t and t'', namely to the event P described by the sentence "Atom A disintegrates at t'." But quantum theory gives absolutely no grounds for predicting the time when a given atom of a radioactive substance will disintegrate; hence there is no potential inductive prediction of P.

I think that Hempel's reply to the present objection should have been the following, which is in essence the one given by Paul Feyerabend in criticism of Hanson.[16] The reply consists in denying the adequacy of the relevant quantum mechanical explanations of micro-events. The argument is that, since we can give no answer to the question "Why did P occur?" that is, to the question "Why did Atom A disintegrate at t'?" we cannot give a satisfactory explanation of P. This reply is not only consistent with the PPP but highly plausible in itself. And the fact that the evaluation corresponds to our intuition may be regarded as a point in favor of the principle. The conclusion is that the PPP survives the second objection.

Self-evidencing Explanations. A third objection to the PPP is one based on the existence of certain explanations which may be called *self-evidencing*. A self-evidencing explanation is one in which the occurrence of the explicandum event provides the only evidence or an essential part of the only evidence for some of the statements contained in the explicans. For example, the explanation of the spectrum of a certain star often consists, among other things, of hypotheses about the chemical constitution of the star and its atmosphere; and the main reason one has for believing in the correctness of the hypotheses is that the spectrum has the features that it does have.

If self-evidencing explanations are to be counterexamples to the PPP, they must be satisfactory but not potentially predictive. Though their reasons differ, both the critic and the supporter of the PPP agree that self-evidencing explanations are not necessarily unsatisfactory; that is, they justify differently their common claim that self-evidencing explanations, qua self-evidencing, are not unsatisfactory. The different justifications are no doubt signs of deeper disagreement and will be explained later, but the only relevant question here, where we are discussing self-evidencing explanations in general, is whether or not they are, qua self-evidencing, potentially predictive.

It is not easy to state the present objection. Scriven argues that in such cases "we have more *data* for explaining than we did for predicting. Hence the former may be certain and the latter not." [17] He also says that "a prediction is by definition such that it *could* be given before the event, but these predictions [the potential predictions which self-evidencing explanations are alleged to be], requiring data from the event, logically could *not* be given before it." [18] What this seems to mean is that self-evidencing explanations are such that the explanatory information could not have been used to predict the explicandum

24

event inferentially, for the simple reason that the information was not, and what is more, could not have been, in these cases, available before the occurrence of the explicandum event that makes it available.

This difficulty will easily dissipate if one takes a close look at what the PPP asserts. It states that all adequate explanations of events are potentially predictive in the sense that they are such that *if* the explanatory information had been available before the occurrence of the event, then it could have been used to inferentially predict the event. The question is then, "Is it true or false of a self-evidencing explanation that if the explanatory information had been available before the occurrence of the event then it could have been used to inferentially predict the event?"

The truth or falsity of this hypothetical depends primarily on the interpretation of the 'if-then'. A reasonably clear and articulated, though somewhat paradoxical view, and the one that the supporter of the PPP advocates, interprets 'if-then' statements like the one under consideration as so-called "material conditionals," that is, as statements which are true when either their antecedents are false or their consequents true or both, and which are false otherwise. Under this interpretation, the hypothetical in question is true by falsity of its antecedent, that is, of the proposition that the explanatory information had been available before the occurrence of the event; and the falsity of the consequent, that is, of the proposition that the explanatory information could have been used to predict the event inferentially is irrelevant. Thus, under a material interpretation of 'if-then' self-evidencing explanations are potentially predictive.

If, on the other hand, the 'if-then' of the definition of potential predictiveness is not construed as a material conditional, it is doubtful that the present objection could even get off the ground. For what the PPP critic would have to show is that the relevant conditional is false. I suppose that for him this would be equivalent to showing that even if the explanatory information of a self-evidencing explanation had been available before the event, it could not have been used to predict the event inferentially. And it seems clear that the critic could not show this and remain consistent with his argument that the explanatory information could not have been used to predict the event because it was not available before the event. But this latter argument shows that a self-evidencing explanation is "nonpredictive" in the sense that it is such that the explanatory information could not have been used to predict the event. Thus it seems to be inconsistent to affirm that self-evidencing explanations are nonpredictive (about which both the critic and the supporter of the PPP presumably agree) and to deny that they are potentially predictive.

At any rate, it is questionable whether self-evidencing explanations are, in fact, even nonpredictive. For one might argue that it is indeed false that the explanatory information was used to predict the event inferentially, but to say this is not to say that the explanatory information lacks any genuine property; in fact, being used to predict an event inferentially is no more a real property of a piece of information than was a genuine property of Descartes's body his ability to doubt the existence of his body (from which, together with a statement of his inability to doubt the existence of his mind, he is alleged to have concluded that his mind was not identical with his body, thus committing some kind of intentional fallacy). But the assertion that the explanatory information could not have been used to predict the event inferentially obviously intends to deny a real property to that explanatory information; hence that assertion must be interpreted as stating that it is logically or physically impossible that the information be inferentially used to predict the event. But, when so interpreted the assertion is clearly false; for example, in the explanation of the star spectrum, it is logically possible to know independently the chemical constitution of the star because the following hypothesis is logically self-consistent: the star in question is a triple star, and triple stars have a known chemical constitution.

The present objection seems to derive strength, however, from the fact that it is not just false that the explanatory information had been available before the event. Given that we are talking about self-evidencing explanations, it is impossible that the information should have been available before the event. That is, the conditional proposition involved is not just counterfactually true, which is all that Hempel admits;[19] it seems to be vacuously true. Thus self-evidencing explanations have to be admitted to be potentially predictive, but at the cost of making them vacuously potentially predictive.

The above considerations, however, present no real problem since, as Hempel himself suggests, the vacuity involved is a rather artificial one and not a natural or logical one.[20] In other words, the class of self-evidencing explanations does not seem to form either a logical or natural kind but only a class brought into being by mere definition. A class of self-evidencing explanations would be a natural kind if the explanations in it were self-evidencing in virtue of the laws of nature;

similarly, it would be a logical kind if the explanations in it were self-evidencing in virtue of the laws of logic. Perhaps one could argue that certain psychological explanations (of human actions) form such a natural kind, but no one has done so. I think it can be argued that explanations of scientific discoveries form such a natural or logical kind, but I have not done so either. At any rate, the PPP would remain true, even though vacuously so for self-evidencing explanations.

Cause-explanations. The fourth objection to the PPP is closely related to the third but is distinct from it. It is based on a type of explanation, which may be called *cause-explanation*, in which an event is explained by stating what caused it. For example, a given case of paresis may be explained by saying that syphilis caused it; or the collapse of a bridge is sometimes explained by saying that fatigue was the cause; or a certain case of skin cancer may be explained by saying that it was due to (its cause was) a high level of radiation exposure.

It is clear, according to Scriven, that these explanations are not potentially predictive because, even if we had known about the presence of the cause, we could not have predicted the effect with either deductive or probable arguments. "In fact, very few syphilitics develop paresis";[21] fatigue is usually not followed by collapse, nor excessive radiation exposure by skin cancer, unless, of course, we arbitrarily and unhelpfully make it part of the definition of "fatigue" and of "high level of radiation exposure" that they are followed by collapse or cancer respectively.

These explanations, Scriven argues, are satisfactory because they are illuminating, informative, and can be satisfactorily supported. These two latter characteristics are closely connected; each explanation is informative because it rules out other causes in the case under investigation,[22] and it can be satisfactorily supported by the elimination of alternatives. The procedure is the following.[23] We infer the explanation, e.g., fatigue caused the bridge to collapse, from the following propositions:

(1) The bridge collapsed.
(2) The only causes of a bridge collapse are excessive load, fatigue, and external damage such as corrosion or explosion.
(3) Neither external damage nor excessive load was present.
(4) Every event has a cause. Therefore, fatigue caused the bridge collapse.

By analogous arguments each of the cause-explanations mentioned can be justified. Moreover, each explanation is informative because the causal statements constituting the explanation are committed to the existence of appropriate links and connections between the cause and the effect.

As I mentioned above, the Hempelian answer to the objection is that these explanations are not really counterinstances because they are either unsatisfactory or potentially predictive. In fact, there are two possibilities. Either we accept Scriven's analysis of those explanations or we provide an alternative analysis. We might accept Scriven's analysis of those explanations as examples of cause-explanations, that is, as examples of events made comprehensible by identification of the cause. But then the explanations would not be satisfactory because they fail to satisfy the general condition of adequacy, that is, the requirement that any rationally acceptable explanation of an event must provide information constituting good grounds for believing that the event did in fact occur; thus cause-explanations such as the ones mentioned above are not satisfactory answers to the question "Why did the event occur?" It is obvious that the supporter of the PPP is not involved in circular reasoning; for he is not saying that the alleged counterexamples are unsatisfactory because they fail to satisfy the PPP, but because they fail to satisfy the principles on which the PPP is grounded. It may of course be that those principles are unacceptable, but what would follow from this is that the PPP need not be accepted because it follows from principles which need not be accepted.

The other possibility is to give a different analysis of the examples given. It is possible to analyze them as examples of self-evidencing explanations. And when we analyze them in this manner they are potentially predictive after all. Let us see what such an analysis would be in the specific instance of a case of paresis. That this alternative analysis presupposes a questionable form of determinism will be examined later. We know that the individual in question has paresis; we also know that the only cause of paresis is syphilis. Thus, accepting the principle that every event has a cause, we conclude that the patient must have or have had syphilis. We also know that only 28 percent of syphilitic patients develop paresis; those who do may be supposed to have the physiological trait of being "paresis-sensitive"; and we are here presupposing the principle that different effects must be due to different causes. At present the only test for the characteristic of being paresis-sensitive is to check whether paresis develops or not;

but it is not hard to imagine that other tests will some day be devised. The individual in question was then paresis-sensitive; his case of syphilis is then being explained by the hypothesis that he is a paresis-sensitive individual who contracted syphilis. And if the information contained in this hypothesis had been available before the occurrence of paresis, then we would have been able to predict its occurrence with deductive certainty or with high probability depending, respectively, on whether we use the universal law that syphilis and paresis-sensitivity always lead to paresis or the statistical law that syphilis and paresis-sensitivity lead to paresis in a high percentage of cases.

In general, then, the explanations of the type mentioned will be analyzed as follows. The explicandum sentence E will be the statement that the event to be explained occurred. The explicans will consist of the following set of three sentences: (1) a statement C asserting that the causal factor was present; (2) a statement $C'(E)$ which asserts that another condition was present, a condition which is absent in the majority of cases when E does not follow C and for the presence of which the occurrence of E happens to provide the only evidence or an essential part of the only evidence; and (3) a general law L stating that if both C and $C'(E)$ occur then E will follow either in every case or in a high percentage of cases. The explanations can thus be given an inferential structure; and they are potentially predictive because, if, before the occurrence of E, we had known that both C and $C'(E)$ were present, knowing the general validity of L, we would have been able to predict with deductive certainty or with high probability that the events to be explained would have occurred.

Since the above explanations, when analyzed in the manner indicated, are potentially predictive, it is actually superfluous for a refutation of the present objection that they should also be satisfactory. However, if the inferential analysis is to avoid obvious evaluative bias, it should not have the consequence that all cause-explanations are unsatisfactory. Hence, the supporter of the PPP must also show that, when analyzed as self-evidencing, cause-explanations are not necessarily inadequate; that is, he has to show that self-evidencing explanations are not inadequate qua self-evidencing. It is true that the supporter has argued above that cause-explanations are usually not satisfactory because they fail to satisfy the general condition of adequacy. But he regards this as a criticism of that analysis of explanation which interprets the relevant causal accounts as cause-explanations. How does he conclude that self-evidencing explanations need not be unsatisfactory?

He argues that, in general, the explanatory problem or question does not even arise unless we presuppose that the explicandum event has occurred. Thus when an argument is being used for explanatory purposes one is not trying to show that the explicandum sentence is true, what one is trying to do is to show that it can be deductively or inductively inferred from the explicans. And in a self-evidencing explanatory argument the inductive or deductive inference is not made any less sound on account of its self-evidencing aspect. In other words, the relationship between the explicans and the explicandum is unaffected by the self-evidencing character.

Moreover, self-evidencing explanations do not involve the explanatory circle of explaining the explicandum "by itself," since the statement of the occurrence of the explicandum is not included in the explicans. What is included in the explicans is a number of propositions all of which are supported by additional evidence besides the occurrence of the explicandum, thus giving empirical content to the explicans. For example, in the inferential analysis of the paresis example, the proposition C that the individual has syphilis and the proposition $C'(E)$ that he is paresis-sensitive are supported not only by his having paresis but also by the fact that the only cause of paresis is syphilis and by the fact that only 28 percent of syphilitic patients develop paresis.

My conclusion is not that the present objection to the PPP has thus been refuted, but that its force is dependent either on the analysis one chooses to give to the alleged counterinstances or on the more or less a priori requirements that one imposes on explanations. For if one accepts either the inferential analysis of explanation and the determinism it presupposes or the general condition of adequacy, then one can show that the objection is not valid; whereas only if one rejects both is the objection successful. In other words, what the present objection really amounts to is the argument that certain satisfactory cause-explanations are not potentially predictive. The reply has been that those explanations in fact are either unsatisfactory, if analyzed as cause-explanations, or potentially predictive, if analyzed as self-evidencing explanations. Thus the critic is actually rejecting both the general condition of adequacy and the inferential analysis of explanation; in so doing the critic may very well be right. But how can such a rejection of those grounds of the PPP be a logically relevant criticism of the PPP? Only if it is presupposed that the PPP follows from those grounds. The objection then has really the form that the PPP need not be accepted because it follows from an unacceptable analysis and unacceptable principles. Thus the present objection presupposes that the PPP is an integral part of the inferential theory of explanation.

The question still remains, "Which analysis is correct?" A full discussion of this question is beyond the scope of my present investigation. I shall, however, examine whether the fact that the inferential analysis presupposes a rather strong and questionable version of determinism presents a serious difficulty for that analysis. Whereas the cause-explanation analysis of the present examples presupposed only the principle that every event has a cause, the self-evidencing explanation analysis also presupposed the principle that different effects are due to different causes. The problem is that, since the latter principle is refuted by quantum theory, it seems the self-evidencing explanation analysis is committed to a falsehood and should thus be rejected. But the self-evidencing analysis is committed at most to the principle that different effects of the kind being discussed are due to different causes of the kind being considered. I say "at most" because this restricted principle is not just presupposed by the self-evidencing analysis but is actually supported by it. For that analysis shows it to be logically and physically possible to claim that in the cases being discussed the different results are due to different conditions. Whereas when quantum mechanical effects are being considered, the assumption that different results are due to different conditions leads to empirical falsehoods. In other words, it is not the case that the self-evidencing explanation analysis should not be given because (predictive) determinism is false; rather, predictive determinism is false because the analysis does not work in all cases, e.g., for quantum mechanical effects. Therefore, given a new or questionable case, what one should do first is to determine whether the self-evidencing analysis works. If it does, that shows quantum mechanical indeterminism to be irrelevant to the case under consideration. The conclusion is that indeterminism presents no difficulties for the self-evidencing explanation analysis of the type of explanation considered in this fourth objection to the PPP—which analysis has disarmed, if not eliminated, the present objection.

Logical Unimportance of the PPP. The objections to the PPP in the next set are unlike the previous ones in that they do not question its truth but its importance, consequences, and use. This criticism depends somewhat on the previous objections in that it is often made on the realization that the PPP is immune to the objections that have been raised against its truth. Having discussed all the objections to the truth of the PPP, we are now in a position to appreciate this kind of criticism.

The first objection here is that the PPP is too general to be of any logical importance because it says nothing particular about explanations. In fact, in any sense in which satisfactory explanations are potentially predictive, all sorts of things, such as descriptions, also are. For example, if we described what happened to Tycho Brahe on the evening of November 11, 1572, by saying that "he saw a star brighter than Venus at her brightest, in a place where no star had been before," [24] it is clear that if we had been in possession of this descriptive information before November 11, 1572, then we could have predicted Tycho's observation.

The PPP is also too general as a characterization of what is Hempel's primary purpose to describe and analyze, namely, scientific explanation. For as we have already observed in part, historical, psychological, medical, and everyday life explanations of events are potentially predictive. Thus in saying that all satisfactory explanations of events are potentially predictive, Hempel is not saying anything particular about scientific explanation. To say something logically interesting about scientific explanation, one must say something which is true of scientific explanation but untrue of, or at least inapplicable to, nonscientific explanations. Nor is this aspect of PPP unimportance rescued by the fact that, in attributing potential predictiveness to satisfactory explanations, the principle has a certain prescriptive character. According to the principle, many quantum mechanical and biological explanations turn out to be unsatisfactory, while certain nonscientific explanations are satisfactory. In other words, the PPP cannot be interpreted as prescribing to nonscientific explanations the features of scientific explanation.

The truth is that, in his attempt to extend the connection between prediction and explanation from some of the natural sciences to all of the empirical sciences, including history, Hempel has receded to a universally true but degenerate form of what I earlier called the predictivist requirement. This is the thesis that all satisfactory explanations of events are such that the explanatory information could have been used to predict the event inferentially. The selectiveness of this requirement is obvious: self-evidencing explanations, including cause-explanations when so analyzed, do not satisfy the requirement, and this is taken to be as much a characterization of their nature, as the fact that explanations in some natural sciences do satisfy it is taken to be a characterization of theirs. The logical interest of the predictivist requirement is equally obvious: the non-applicability of the requirement can be taken as a defining characteristic of historical explanation; and all that one is using here is the uncontroversial idea that history is the study of the past, with the consequence that prediction of the past excludes itself from that study.

PPP Philosophically Misleading. In claiming that all satisfactory explanations of events are potentially predictive, the PPP insinuates that all satisfactory explanations are predictive after all, though in a special sense. That is, it suggests that all satisfactory explanations are a special kind of predictions, "potential predictions." That this suggestion is real is shown by the fact that Hempel himself often states the PPP by saying explicitly that satisfactory explanations of events are potential predictions.[25] However, potential predictions are no more predictions than imaginary bank accounts are a special kind of bank accounts. For the proposition that all satisfactory explanations of events are potential predictions is supposed to mean that satisfactory explanations of events are such that if the explanatory information had been available and taken into account before the event, then it could have served to predict the event inferentially. A potential prediction is then an argument or an account of an event which, if it had been available and taken into consideration before the event, could have served inferentially to predict the event. Thus it should be noted that a potential prediction is not the same as what might be called a hypothetical or conditional prediction, i.e., a statement of the form "If A happens (or if A happens at t), B will occur (or B will occurt at t')." Indeed such hypothetical predictions become potential ones only if the condition is given; that is, if we regard the above conditional proposition as a hypothetical prediction of B, then that proposition cannot serve to infer that B will occur unless we also know that A has happened or can infer that it has happened or will happen.

According to Hempel's point of view then, a potential prediction is a special case of a prediction, i.e., a special case of an account that can serve to infer the event before its occurrence. Let us ask ourselves what other kind of prediction a potential prediction is being contrasted to. If we have a genuine special case, then there should be a complement class of special predictions, that is, a class which together with potential predictions, would constitute the whole class of predictions. What can such a complement class be? Could such a complement class be the class of all nonpotential predictions, that is, of all accounts of events of which it is not true to say that if they had been available and taken into consideration before the events they could have served to infer the events before their occurrence? But, at least for Hempel, such a class would consist of those accounts which had been available and taken into consideration before the events but were not sufficient to infer the events before their occurrence. But an account which is not sufficient to infer an event before its occurrence is either a false prediction or not a prediction at all. Therefore, the class of predictions complement to potential ones would be a subset of false predictions or of nonpredictions. Such a subset is surely the wrong class.

Suppose we say that the complement class consists of those accounts of events that can serve to infer the events before their occurrence but are such that they were either unavailable or not taken into consideration before the event. But arguments which are either unavailable or not taken into consideration before an event are not predictions of the event either. Therefore, we have not yet identified the complement class.

Potential predictions are not a subclass of genuine predictions at all. Thus to state the present consequence of the general condition of adequacy as a connection between the concepts of explanation and prediction is philosophically misleading; what should be said is that satisfactory explanations of events are actually *retrodictive*.

I have pointed out above the misleading character of the PPP due to the expression *potential prediction*. I shall now point out another misleading aspect, due to the term *potential*. In fact the PPP suggests that, though not every satisfactorily explicable event is actually predicted, it is nevertheless predictable. This happens through the following series of conscious or unconscious transformations. The statement that every satisfactory explanation of an event is a potential prediction of that event becomes, when expressed with verbs in the passive voice, the assertion that any event is potentially predicted by any satisfactory explanation of it. This latter statement turns into the proposition that any satisfactorily explained event is thereby potentially predicted; next, that any satisfactorily explicable event is thereby potentially predicted; and finally, any satisfactorily explicable event is thereby predictable; that is, contrapositively expressed, if an event is not predictable, it is not satisfactorily explicable.

Since the PPP is irrefutably true, and since the proposition that unless an event is predictable it is not satisfactorily explicable is highly implausible, the PPP misleads one to think that that implausible statement is justified. The implausibility of this claim is obvious: what it really claims is that if the occurrence of an event cannot be inferred at some appropriate time before the event, then its occurrence cannot be inferred any time after the event. But to say this is to say that our capacity to make inferences does not improve with the passage of time; that is, we do not learn from experience or from any other source; that is, there is no growth of knowledge. And this latter claim I take to be empirically false.

But why cannot "potentially predicted" be equated with "predictable?" Simply because the former means "inferred from an account *A* which can serve to predict the event if *A* is available and taken into consideration before its occurrence"; whereas "predictable" means "inferrable before its occurrence."

Application to the History of Science

So far we have seen that the PPP is true but logically unimportant and philosophically misleading. From the point of view of the history-of-science explanation it is also logically vacuous, methodologically absurd, and methodologically misleading.

Logical Vacuity. The PPP is logically vacuous because explanations of scientific discoveries, perhaps the most important class of history-of-science explanations, are vacuously potentially predictive. That is to say, they are such that if the explanatory information had been available and taken into account before the discovery, then it could have been used or would have been sufficient to predict the discovery inferentially, merely because that information is not both available and taken into account before a discovery. In order to show that satisfactory explanations of scientific discoveries are vacuously potentially predictive, it must be demonstrated that satisfactory explanations of scientific discoveries are such that the explanatory information is either not available or not taken into account before the discovery. At the most general level, a scientific discovery is a historical event or development whereby new knowledge comes into being. If, to explain satisfactorily why this new knowledge came into being, we have to provide information showing that the new knowledge did in fact come into being, then clearly the explanatory information could not have been both available and taken into account before the discovery. Otherwise the new knowledge would have come into being before it actually did or it would have come into being differently from the way it did; that is, it would have come into being as a result of and in the process of trying to predict inferentially that it would, which is not the way scientific discoveries are ever made, or could normally be made. Scientific discoveries could not normally be made by attempting to predict them inferentially because there is a conceptual distinction between a scientific discovery and an inferential prediction of a scientific discovery.

Since the unavailability of the explanatory information before the event is due, in the case of scientific discoveries, to the nature of scientific discovery, it is logically impossible that the explanatory information be available and taken into account before the event. Thus scientific discoveries are indeed potentially predictive, but vacuously so. As I mentioned before, Hempel recognizes that such a fact would present this problem for the PPP. What is the problem? It is that to claim the explanation of scientific discoveries to be potentially predictive is not even to say something tautological about such explanations, as it is perhaps the case in other fields; it is literally to say nothing about them. In other words, the thesis that satisfactory explanations in physics are potentially predictive says nothing particular about explanation in physics—and this is bad enough. But the thesis that the historical explanation of scientific discoveries is potentially predictive says nothing at all about history-of-science explanation. Hence, if one wants to describe certain predictive aspects of history-of-science explanation, or even prescribe them to it, one cannot do it in terms of the PPP.

Methodological Absurdity of the PPP. My argument showing the logical vacuity of the PPP assumed that there were or could be "satisfactory" explanations of scientific discoveries. Although this may very well be true, it is a consequence of the PPP that no explanation of scientific discoveries can be satisfactory. Suppose, in fact, that we state the PPP in the way Hempel sometimes does, namely, by saying that all satisfactory explanations of events are potential predictions. I think it can be shown that there are no satisfactory explanations of scientific discoveries because there are no potential predictions of scientific discoveries, and there are no such potential predictions because the notion of the potential prediction of a scientific discovery is logically inconsistent, as I shall demonstrate.

Let d be a scientific discovery and assume that we describe it by the sentence "At t, B discovered that p," where B is the name of an individual scientist and p a proposition stating, e.g., a law of nature.

Let $A(d)$ be an argument constituting a potential prediction of d. Then $A(d)$ is such that, if it had been available before t, it would have been sufficient to predict the occurrence of d inferentially. It follows from this that the occurrence of d is a consequence of $A(d)$.

But if $A(d)$ has been sufficient to predict the occurrence of d inferentially, it would have been sufficient inferentially to predict that at t, B would discover that p. But inferentially to predict that at t, B would discover that p is to have conclusive or good reasons for asserting that at t, B will discover that p, i.e., it is to infer before t that at t, B will discover that p, and consequently to infer that p is true, since the sentence "p is true" is an obvious consequence of the sentence "At t, B will discover that p." Therefore, if $A(d)$ had been available before t, it would have been sufficient to infer that p is true.

But to infer before t that p is true is to discover that p. Therefore, if $A(d)$ had been available before t, it would have been sufficient to discover that p. But if $A(d)$ is sufficient to discover that p before t, then it is sufficient to infer the prediction at t, B will not discover that p, since no scientific discovery can be made of what has already been discovered. Therefore, if $A(d)$ had been available before t, it would have been sufficient to predict inferentially that at t, B would not discover that p, that is it would have been sufficient to predict inferentially the nonoccurrence of d. It follows from this that the non-occurrence of d is a consequence of $A(d)$.

Therefore, both the occurrence and the non-occurrence of d are consequences of $A(d)$; therefore, $A(d)$ is logically inconsistent.

In the preceding argument the term *discovered* appeared in the description of the event. In order to show that the contradiction which we derived is verbally independent of this term one could consider the case when a discovery is described by the sentence "N was the first to prove that p," e.g., Newton was the first to prove that an elliptical planetary orbit is a consequence of an inverse-square force field. Or one could consider the case when the discovery is described by a sentence of the form "L gave a new explanation of e," e.g., Lavoisier gave a new explanation of the weight augmentation effect. Analogous contradictions would follow.

It should be noticed that the argument does not show that scientific discoveries are unpredictable. Indeed this would have been both insufficient and unnecessary for my purposes. It would have been unnecessary because all I had to show was that scientific discoveries are potentially unpredicted, and, as we saw above, "potentially unpredicted" is not the same as "unpredictable." It would also have been insufficient because, just as inferability after the event does not guarantee inferability before the event, so non-inferability before the event does not guarantee non-inferability after the event. Indeed the history of science gives paradigm examples of unpredictable but post-inferable discoveries, such as the discoveries of oxygen and X-rays, which exhibit the fundamental inadequacy of a previous general theory. The discovery of phenomena contradicting the theory is intrinsically excluded by the theory itself; but when the discovery has been assimilated into a new theory one can often argue that the discovery *was bound to* (or was likely to) take place in view of the cognitive state and experimental conditions in which the discoverer found himself. Such an argument, if it were not self-contradictory to say so, "potentially predicts" the discovery.

The above argument may lead the supporter of the PPP to claim that satisfactory explanations of scientific discoveries are not possible, but it leads the historian of science and the methodologist of history of science to the conclusion that the PPP is absurd; the above argument may then be interpreted as a *reductio ad absurdum*.

PPP Methodologically Misleading. Besides being logically vacuous and methodologically absurd as concerns history-of-science explanation, the PPP is methodologically misleading. For to speak of predictiveness, albeit "potential," when one ought to speak of *un*predictiveness (or non-predictiveness or anti-predictiveness) is to make the historian of science think of something worse than irrelevant and absurd; it is to lead him to think of the opposite of what he should. In fact, if one wants to find a connection between explanation and prediction in the historiography of science, that connection is the following: explanations of scientific discoveries are such that the explanatory information could have been used to predict the *non*-occurrence of the discovery. The argument would be as follows.

One aspect of virtually all scientific discoveries is that (at least) one individual infers a proposition p for the first time in human history. In other words, any historically complete description of a scientific discovery would have to include, no doubt, among other things, a description of the form "*i* was the first to infer that *p*," where *i* stands for the name of a scientist and *p* for a proposition expressing the content of the discovery. One example would be the description that Kepler was the first to infer that the orbit of Mars is elliptical. Suppose we take this Keplerian discovery as our explicandum. The explanation of this aspect of Kepler's discovery might be of the following form: Kepler inferred that the orbit of Mars is elliptical because $R_1, \ldots, R_n$ and he was the first to draw the inference because he had followed Tycho Brahe in abandoning crystalline spheres which had made ellipses impossible;[26] the R_i's in this explanation would express Kepler's reasons, be they theoretical, factual, observational or metaphysical. The explanation consists of two parts: an account of how Kepler arrived at his conclusion and an account of why no one else did before him. These two analytical structures need a great deal of investigation; but our main purpose here is to exhibit the unpredictiveness of such explanations. In order to facilitate this, we shall give the explanation an inferential structure, as Hempel would. Hempel's model of the explanation of human actions by motivating reasons is the following:

> The answer to the question "Why did *A* do *x*?" takes the following form:
> A was in a situation of type *C* [i.e., in conditions $C_1, \ldots,$ C_n].
> (Schema R)   A was a rational agent.
> In a situation of type *C*, any rational agent will do *x*.
> Therefore, *A* did *x*.[27]

Therefore there is little doubt that the 'how' account should be given a structure like the following:

(*C'*) Kepler believed that $R_1, \ldots, R_n$ are true.
(*C''*) Kepler was a rational thinker.
(L_1) Any rational thinker who knows that $R_1, \ldots, R_n$ infers that *p*.
Therefore Kepler inferred that *p*.

I am not saying that it would be easy to establish such an argument, though this would depend in part on whether the 'how' account was acceptable. I am only claiming that this is the inferentialist analysis of that account. In the same vein, the inferentialist reconstruction of the account of why no one inferred p before Kepler would be the following:

(1) Every rational thinker before Kepler believed in the theory of crystalline spheres.
(2) This theory makes elliptical orbits impossible.
(3) Every rational thinker before Kepler failed to infer that p.
(4) Therefore Kepler was the first to infer that p.

Consistency will require that the combination of these two arguments should depart somewhat from the principle that laws with essential occurrences of individual names are unacceptable for explanatory purposes. In fact, (L_1) will have to be changed to the assertion, which is still importantly general, that any rational thinker after (and including) Kepler who knows that $R_1, \ldots, R_n$ infers that p. This change is also necessary since (L_1) is supposed to exclude any evaluative connotation. As it is obvious, the change increases the plausibility of that first reconstruction since the new generalization is much more plausible. The full inferential reconstruction of this aspect of Kepler's discovery is then the following:

(C') Kepler knew that $R_1, \ldots, R_n$.
(C'') Kepler was a rational thinker.
(L) Every rational thinker before Kepler failed to infer p from $R_1, \ldots, R_n$ and does so after (and including) him.
Therefore Kepler was the first to infer that p.

Let us now investigate the "predictive" aspect of this explanation. It should indeed be admitted at the outset that the explanation is "potentially predictive," counterfactually so. But why concentrate on such a pseudopredictive aspect? The really relevant question is the following: With respect to the explicandum event, what predictive property does the explanatory information have? That is, what predictive inference does the explanatory information allow with respect to the explicandum event? In other words, prior to Kepler's own inference, what inference could have been drawn concerning Kepler from the explanatory information? The only information contained in the explanation and relevant to such an inference warrants the conclusion that Kepler will not infer that p and a fortiori will not be the first to do so. Thus the explanatory information is such that it could have been used to predict the non-occurrence of the discovery, which may be summarized by saying that the explanation is anti-predictive. The PPP is thus methodologically misleading, while its nondegenerate form, the predictivist requirement, is also objectionable since it is not just inapplicable but applicable in reverse. My general conclusion is that the predictive aspects of the explanation of scientific discoveries are anomalous and present great difficulties for the PPP and some even for the predictivist requirement.

chapter **3**

The Explanation of Discoveries:
Law-Coverability

THE PPP is not the only principle that encounters difficulties when applied to history-of-science explanation. Problems arise also in the application of a second principle, which attributes a second general feature to explanation, one that constitutes a connection between the concept of explanation and that of generalization, regularity, or law. The principle can be stated as follows: every satisfactory causal explanation contains, when fully stated, at least one nonredundant universal or statistical generalization.

The two basic ideas behind this principle, and from which one can construct an a priori and an a posteriori argument in its support, are the following. First, the explanation of phenomena like solar and lunar eclipses as well as the answer to questions like "Why did Apollo 8 acquire a speed of about 24,000 mph in its return to earth?" make reference to, and would include if fully stated, generalizations such as the law of gravitation and the laws of motion. The second idea is that if one claims that a certain condition or number of conditions *c* cause a given phenomenon or event *e*, then so it is alleged, one is implicitly claiming not only that *c* preceded or was simultaneous with *e* but also that whenever a condition of the type *c* occurs, so does a condition of the type *e*.

Since I am here primarily concerned with the difficulties that history-of-science explanation presents for the principle stated above and not with its truth or falsity, I shall not elaborate upon these two ideas but rather upon the statement of the principle itself. First, it should be noticed that the principle mentions only causal explanations and is thus not meant to apply to other types of explanation.

Second, the principle speaks of generalizations which are either universal or statistical. In this context, universal generalizations are supposed to be statements like the laws of motion or the gas laws; statistical generalizations are statements like the half-life laws for radioactive substances. The principle was weakened to allow for statistical generalizations partly in order to interpret as possibly satisfactory explanations such historical assertions as: William the Conqueror never invaded Scotland because he had no desire for the lands of the Scottish nobles and he secured his northern borders by defeating the King of Scotland in battle and exacting homage.[1] In such cases, according to the present interpretation, no difficulty results from the lack of a universally valid law, at least a known law, correlating lack of desire to expand and sense of security with non-invasion. No difficulty follows because the following statistical generalization is true: if a king has no desire for territorial expansion and feels secure within his borders, he is very likely not to invade neighboring lands.

Third, the principle speaks of nonredundant generalizations. In this context a generalization is nonredundant if it needs to be used for the deductive or inductive inference of the explicandum from the explicans. If the explicans still implies the explicandum, even when the generalization is not part of the explicans, then the generalization is redundant. For brevity, we may call a nonredundant universal or statistical generalization contained in a fully stated explanation a *covering law* for the explanation.

Fourth, the principle speaks of "at least one" covering law. In so doing it implies that the general case is one in which there are several laws covering the given explanation. The principle thus avoids the criticism sometimes made of other versions of it. The criticism is that by requiring an explanation to be covered by a single law, the principle makes the satisfactoriness of a historical explanation, for example, depend on a generalization which is not a true generalization, but simply an ad hoc generalization of the individual case under investigation.

Finally, and most importantly, the principle includes the condition "when fully stated." It does not say that every satisfactory causal explanation contains some covering law, but that it does so if a certain condition is met, and the condition is that it be fully stated. One might say, in other words, that it is not the explanation itself that is law-covered, i.e., that contains the covering law, but the "full statement" of the explanation. The present principle then amounts to saying that the full statement of any satisfactory causal explanation is law-covered. The structure of this principle is that of a conditional proposition with a conditional as its consequent, as exhibited in the following statement: If something is a satisfactory causal explanation, then if it is fully stated, it will be law-covered. For brevity we may introduce the suggestive term *law-coverable* as meaning "law-covered when fully stated," and thus say that the principle under consideration states that every satisfactory causal explanation is law-coverable. This principle will be called the *Law-Coverability Principle for Explanations* (LCP for short) and should of course be distinguished from the statement that all satisfactory explanations are law-*covered*.

There are further clarifications to be made in answer to two questions. The first asks what is or could be meant by a full statement of an explanation. The second asks whether the term *fully stated* carries with it a methodological evaluation to the effect that it is better for an explanation to be fully stated than not to be. The term *fully stated* will here be taken as being evaluatively neutral, so that an explanation will not necessarily be made unsatisfactory if it is not fully stated, and indeed may be unsatisfactory, e.g., boring, if fully stated. And an explanation will be said to be fully stated when one explicitly states not only the causal conditions by reference to which the explicandum is being explained, but also the universal or statistical generalizations in virtue of which those conditions are producing the explicandum. In other words, to fully state an explanation, which is not already fully stated, one has to give it an inferential structure. The LCP thus says, in effect, that every satisfactory causal explanation, when inferentially analyzed, contains a covering law. Using the notion of the explicans, which is already an analytical notion so that the explicans may have to be constructed if it is not explicit, the LCP can also be stated as follows: Every satisfactory causal explanation contains a covering law in its explicans.

A fully stated explanation is then an argument of the following form:

$$\left.\begin{array}{l} C_1, C_2, \ldots, C_n \\ L_1, L_2, \ldots, L_k \end{array}\right\} \text{ Explicans}$$

$$\overline{E} \qquad \qquad \text{Explicandum sentence.}$$

Here the C_i's are the particular facts and conditions by reference to which the explicandum is being explained, the L_i's are universal or statistical generalizations in accordance with which those conditions are being alleged to produce the explicandum, and the inference from the explicans to the explicandum is either deductively certain or inductively probable.

Thus one begins to suspect that the LCP has no more logical interest than the PPP. For in the PPP case the interesting question turned out to be "when can the explanatory information be available and be taken into account before the explicandum event and when can it not?" Similarly, for the LCP the interesting question can be expected to be whether the explanations of a particular field can be fully stated or not, and if they can, how fully should they be stated. And consequently if one is interested in differences more than similarities, one may say that scientific explanations are law-covered, that is they can be and in fact are fully stated, whereas historical explanations are not law-covered, that is, they cannot be fully stated and in fact are not. And this characterization besides being independently justifiable[2] is also intuitively plausible since it corresponds to the idea that history is interested in the particular, science in the general.

Unless one interprets the relation between the property of being fully stated and that of being law-covered in a viciously circular manner, meaning if one attaches some logical content to the LCP, it seems then to be false, as I shall proceed to show. One argument is implicit in our discussion of the explanation of Kepler's discovery of the elliptical orbit of Mars. Let us assume that there are satisfactory explanations of discoveries which are fully stated. It is not easy to imagine on what grounds the explanation of Kepler's discovery we gave could be rejected by a supporter of the PPP or LCP. If we take that example as a satisfactory and fully stated explanation, the important question to consider is whether the explanation is law-covered or not. It will be remembered that the law needed in that explanation was the following statement: Every rational thinker before Kepler failed to infer p from $R_1, \ldots, R_n$ and does so after (and including) him. This statement, as the supporter of the LCP would surely agree, is not a genuine generalization on account of its reference to Kepler. Since the statement does not even have the form of a law, it cannot be said the present explanation is law-covered; we must say it is not. Thus there are satisfactory and fully stated explanations which are not law-covered.

The second argument to show that some satisfactory and fully stated history-of-science explanations are not law-covered is the following. In history of science one will often want to know why a given scientist reached a certain conclusion, e.g., why Leverrier concluded that there was a transuranic planet, or why Kepler concluded that there existed just six planets. "Sketchy" statements of such explanations might be the following. Leverrier concluded that there was a transuranic planet because there were perturbations in the orbit of Uranus and the almost exact values of such perturbations could be derived from the laws of motion, the law of gravitation and the hypothesis of the existence of a planet beyond Uranus. Kepler concluded that there were just six planets because the nature of three-dimensional space was such that there exist exactly five perfect solids and this number is the number of intervals between the planets whose orbits will therefore lie in spheres circumscribing each of the perfect solids. These explanations are of the form "S concluded that q because $p_1, \ldots, p_n$" and they will become more and more fully stated and less sketchy as the number of p_1's increases. From the point of view of the LCP the full statement of such explanations takes the following form:

$$
\text{(I)} \quad \frac{\begin{array}{l} C_1, \ldots, C_n \\ L_1, \ldots, L_k \end{array}}{E.}
$$

The explicandum sentence will be "S concluded that q." Each C_i can be interpreted as the cognitive condition or intellectual state described by "S believed that p_i." The explanation under discussion will take the following form:

        S believed that $p_1, \ldots,$ and p_n

(II) $L_1, \ldots, L_k$
       $\overline{\text{S concluded that } q.}$

The present problem is to determine whether these L_i's are covering laws. I shall argue that they are not. Without begging the issue of whether the behavior of the scientist in question was rational or not, we may say that there are two possibilities: either we have an instance of rational behavior or correct thinking, that is to say that q follows deductively or inductively from the p_i's, or we have an instance of nonrational or irrational behavior or incorrect thinking, that is, that q does not follow from the p_i's. I am not saying, of course, that it would be easy to establish either alternative, but only that they are the only two possible.

In the case of the first alternative the supporter of the LCP would probably complete Schema II by explicitly stating the generalizations descriptive of rational agents and thinkers who infer q from the p_i's. Without loss of generality for my present argument one can combine all the individual L_i's into one generalization L asserting that all rational thinkers who believe that $p_1, \ldots,$ and p_n conclude (or are highly likely to conclude) that q. We thus get the following schema:

     (1) S believed that $p_1, \ldots,$ and p_n.
     (2) S was rational.
(III) (3) All rational thinkers who believe that $p_1, \ldots,$ and p_n conclude (or are highly likely to conclude) that q.
     (4) Therefore S concluded that q.

The problem with this schema is due to (3). Though it is not disqualified from being a law as the generalization in our previous argument was, it turns out not to be a covering law because it actually is redundant in the derivation of the explicandum (4). In fact, since we have already assumed that q follows from the p_i's all we need to derive our explicandum conclusion, besides the proposition that S believed the p_i's, is the fact that S was rational. It is not possible for (4) to be false and (1) and (2) to be true (assuming of course that q follows from the p_i's).

In fact, the generalization (3) is as useless or as useful as a valid principle of inference is in deriving a certain conclusion from certain premises. For example, consider the following argument:

(A) All men are mortal.
(B) Socrates is a man.
(Z) Therefore Socrates is mortal.

It is useless to add as an extra premise the following statement which is a generalization about the kind of proposition that can be inferred from propositions of given forms: (C) propositions of the form "S is M" (i.e., "Socrates is mortal") follow from the conjunction of propositions of the form "All H are M" and "S is H" (i.c., propositions like "All men are mortal" and "Socrates is a man," respectively). As C indeed states (Z) follows from (A) and (B) without the need of (C) as an additional premise in the argument.

The analogy will become more obvious if we give simple interpretations to q and the p_i's which exemplify the kind of case being discussed, i.e., when q follows from the p_i's. Let p_1 be 'P' and p_2 "If P then Q," $n = 2$, and q be 'Q'. Consider the following inference:

(IV)
    (1′) S believed that P and that if P then Q.
    (2′) S was rational.
    (3′) S concluded that Q.

All I am saying is that (3′) can be derived from (1′) and (2′) without the help of the pseudocovering law that any rational thinker who believes that P and that if P then Q concludes that Q. My conclusion is that there may be satisfactory and fully stated explanations of simple historical facts of the form "S concluded that p" which are not law-covered, if we use the rational behavior model which is probably the one that would be needed in cases like Leverrier's conclusion about a transuranic planet.

But if we are faced with a case of idosyncratic behavior or thinking such as Kepler's derivation of the number of planets, then a full statement of the explanation could not include any covering law, since it could not include the appropriate generalization, but rather a specific fact about the individual involved, such as (2″) in the following schema:

(V)
    (1″) S believed that $p_1, \ldots,$ and p_n.
    (2″) S reasoned that since $p_1, \ldots,$ and p_n then q.
    (3″) S concluded that q.

Nor could it be said that in such cases the justification for (2″) would include the appropriate kind of generalization; the idiosyncratic nature of the case precludes this. The justification would probably be based on either a judgment formed after examining S's writing or S's own explicit *saying* that he concluded that q because $p_1, \ldots,$ and p_n. Therefore in any case, not all satisfactory and fully stated explanations are law-covered.

Of course, the supporter of the LCP is in some sense "free" to claim that explanation schemata like III and V are either not fully stated or not satisfactory. This would safeguard the truth of the LCP. But to take this step would be to take the first step toward vicious circularity and vacuity, the methodological undesirability of which, as we shall soon explore, would be too high a price for its truth. That is, what we have considered so far is the possibility that the LCP might be nonvacuously true; and we have argued that this is not the case, by finding examples of history-of-science explanations which, at least from the point of view of the supporter of the LCP, seemed to be good candidates for being both satisfactory and fully stated but which were not, for one reason or another, law-covered. But perhaps we considered the wrong possibility, perhaps the LCP is true after all, and it just so happens that there can be no satisfactory explanations of the above-mentioned aspect of scientific discoveries. To show that the truth of the LCP can be retained only at the cost of making it methodologically absurd, I shall now argue that the notion of a law-coverable account of a scientific discovery is self-contradictory.

Let us assume that there is an account of a scientific discovery with the property that if it is fully stated then it is law-covered. Sometimes the historian of science will want to describe a certain development of scientific knowledge by saying that a given scientist gave a new interpretation of certain old and well-known data. For example, Copernicus gave a new interpretation of celestial motions, Lavoisier of chemical phenomena. If the LCP is to be a principle for explanations, it ought not to legislate about or exclude certain descriptions. Let us consider the explanation X of the event described by the sentence "Lavoisier gave a new interpretation I of chemical phenomena P_c." Suppose that X is law-coverable. Assume now X to be fully stated; X is then also law-covered, i.e., its explicans contains at least one nonredundant generalization L; that is to say, L is such as to enable us to show that, given the particular conditions also mentioned in X, Lavoisier's giving his interpretation of chemical phenomena follows in accordance with

L. But if this is so, if we are in possession of a generalization in accordance with which Lavoisier gave his interpretation, then we have shown that Lavoisier's interpretation was not new, which is inconsistent with our initial description.

It could be objected that all we would have shown is that Lavoisier did not give a really new interpretation, but only an interpretation which was thought to be new. The problem still remains, however; for it is doubtful that anyone can simultaneously adopt both points of view, the philosophical one from which the interpretation is not new and the historical one from which it is. If such double-think is not possible, then when the event is historically described, the explanation of that event could not be law-coverable; that is, it could not be and remain the explanation of the event it is supposed to be explaining. The fact that the explanation of the event described differently, which is to say of *another* event, is law-coverable is irrelevant. At any rate, even if the above mentioned double-think is possible, it seems obvious that the historian of science, almost by definition, must take the point of view from which Lavoisier's interpretation *was* a new one. Since an element of novelty is present in all scientific discoveries, and since that aspect cannot be law-coverable, there can be no satisfactory explanations of scientific discoveries. The LCP is thus true, though vacuously so and methodologically absurd from the point of view of the historiography of science.

But perhaps the wrong choice was made once again: perhaps the LCP is neither false nor vacuously true. I believe, however, that the possibility left is no more redeeming; I shall now argue, in fact, that if the LCP is nonvacuously true, it must be because satisfactory explanations of scientific discoveries are *vacuously law-coverable,* i.e., have the property that if they were fully stated they would be law-covered simply because they cannot be fully stated. If we describe a given scientific discovery by the sentence "S discovered that *p,*" the full statement of its satisfactory explanation would be an argument with the following structure:

$$C_1, \ldots, C_n$$
$$\underline{L_1, \ldots, L_n}$$
$$\text{S discovered that } p,$$

where the conclusion follows deductively or inductively from the C_i's and the L_i's but not from the C_i's alone. The argument is said to deductively or inductively subsume the explicandum discovery under certain laws; it shows that the explicandum discovery is a special case of the laws of discovery. Thus in order to state fully a satisfactory explanation one must be in possession of the "laws of discovery." But it is a simple sociological fact about historians of science and methodologists that no such laws of discovery are known to them. Hence no satisfactory explanation of a scientific discovery can be fully stated by historians or methodologists. Hence, if satisfactory explanations of discoveries are law-coverable after all, that fact would be of overwhelming methodological unimportance.

Moreover, if we follow Karl Popper[3] in regarding temporal invariance as part of the concept of law, it seems logically impossible that there should be known laws of discoveries. For the knowledge of such laws would mean the end of the growth of scientific knowledge, and consequently of discoveries; and those alleged laws could at best be inductive generalizations about past discoveries.

Besides being methodologically absurd and methodologically unimportant, if true, the LCP would be logically uninteresting; for as in the case of the PPP, if all satisfactory causal explanations are law-coverable, this tells us nothing at all about history-of-science explanation, and nothing particular about "scientific" explanation or explanation in the natural sciences. Content has been sacrificed for universality. If one wants specific characterizations of various fields one must look at the differences and not at the similarities. The principle suggested by the LCP and of which it is a degenerate form is what may be called the covering-law principle that all satisfactory causal explanations are law-covered. In terms of this principle one may define another logical difference between scientific and historical explanations: it is true of the former, false of the latter;[4] satisfactory historical explanations are not law-covered. And the intuitive plausibility of this idea is obvious from the fact that science is interested in the general, history in the particular. In terms of the LCP the interesting question is not its truth but whether, if true, the condition mentioned in the antecedent of its consequent, i.e., that the explanation be fully stated, is or can be fulfilled. Analogously in the PPP case the important question was "When can the explanatory information be available and taken into account before the event and when can it not?"

Finally, the LCP is methodologically misleading, since in implying that satisfactory explanations of scientific discoveries are law-coverable it suggests that they are law-covered. And the characteristic of being law-covered ought not to be just absent from explanations of discoveries, but its opposite should be present. That is, it is more nearly true to say that explanations of scientific discoveries are law-free, i.e., explanations of scientific discoveries should avoid the use of covering laws. The reason for this is simple and basic. To say that satisfactory explanations of scientific discoveries are law-covered is to suggest that the cognitive behavior involved in scientific behavior is rule-governed behavior; but rule-governed behavior is not creative, whereas it is of the essence of discovering that it be creative: hence if scientific discoveries are to involve creative behavior then it is necessary to explain them as instances of rule-breaking behavior; that is, the explanations of them should be law-free. My general conclusion is that history-of-science explanation presents fundamental difficulties for a second philosophical principle for explanations, or to express it differently, the concept of the explanation of a scientific discovery is anomalous from the point of view of that principle.

chapter 4

A Third Anomaly and a Reconsideration

Explanations of Discoveries as Answers to 'Why-not' Questions

A THIRD ANOMALOUS aspect of history-of-science explanation can be seen best by trying to apply the idea that explanations are answers to 'why' questions. This idea seems so obvious and basic that it may be pretentious to formulate a principle out of it, all the more so since what we are dealing with here is not so much a property that satisfactory explanations have but rather a defining characteristic of explanations: explanations just are answers to 'why' questions. Of course, the 'why' question need not be explicitly asked; for something to be an explanation it suffices that it be an answer to a question that could be asked. This possibility will be explicitly taken into account by saying that explanations are answers to actual or potential 'why'

53

questions. Since I think the thesis that explanations are answers to actual or potential 'why' questions is dubious,[1] I shall call it the *Why Principle for Explanations* (the WP for short). Though the WP is true of many explanations it is not true of all.

My main concern, however, is to show that the WP is misleading for history-of-science explanation. In fact the WP leads one to believe, as it has led Hempel,[2] that the explanation of an event is the answer to the question "Why did the event occur?" And the understanding is that what is being asked is why the event occurred at the time at which it did.

In applying the WP to history-of-science explanation and to the explanation of scientific discoveries in particular, one is led to the idea that insofar as scientific discoveries are historical events, to explain them one has to answer the question of why they occurred at the time at which they did.

But, as I shall illustrate below, where one is concerned with scientific discoveries there is really no problem about the 'why' of their occurrence at the time at which they did occur (though there may be a problem of 'how' they did occur); the 'why' problem is why they did not occur before they actually did, that is, insofar as there is an explanatory problem about their occurrence at the time at which they did occur, it is a problem about the 'why-not'. The explanation of a scientific discovery may be said, from the present point of view, to be an answer to the question "Why did the discovery not occur before?" In slogan-like terms one may then say that explanations of scientific discoveries are answers to 'why-not' questions.

Is this a problem for the WP? Are not 'why-not' questions a special case of 'why' questions? In some sense they are. But even from a semantic point of view, speaking in this manner is like saying that one person is attracted to another even when they repel each other because each is negatively attracted to the other. Moreover, the WP is indirectly philosophically misleading since it leads one, as it has led Hempel, to formulate the PPP and the LCP. Most importantly it is practically misleading since it makes the historian of science think that he has to find the causes or reasons (i.e., the 'why') rather than the obstacles (the 'why-not') of scientific discoveries.

The 'why-not' aspect of the explanation of scientific discoveries is a necessary consequence of growth of scientific knowledge. Consider the discovery by Archimedes of the principle that bears his name.[3] Could one seriously ask the question "Why did Archimedes discover the principle at the time at which he did?" Only if one were ignorant of Archimedes' Principle. (And even in this case the questioner would not even know what he was asking.) But if one knows and understands the principle, then the principle appears so obvious that the 'why' question does not arise. The question that does arise is either "Why did Archimedes not discover the principle before?" or more likely "Why did the principle have to wait for Archimedes to be discovered?"

There may of course be disagreement about the answer to the latter question, and this disagreement may not be just about which obstacle is responsible but even about the kind of obstacle. Joseph Agassi, for example, thinks that the obstacle was Aristotle's theory of hydrostatics.[4] But the disagreement is not and, from the nature of the case, could not be about which question the explanation ought to answer.

Another good example illustrating the 'why-not' aspect of the explanation of scientific discoveries is Torricelli's discovery of the vacuum. The only 'why' question here would be "Why wasn't the vacuum discovered centuries before Torricelli when the limitations of the suction pump had been discovered?" Similarly, the explanation of the discovery of the laws of falling bodies would have to answer questions like "Why wasn't the parabolic trajectory of projectiles discovered before Galileo, e.g., by the ballistic technicians who constructed charts of those trajectories?"

Our conclusion is that the Why Principle for Explanations, just like the two previous principles, cannot be accepted, at least not without a metamorphosis.

Transition to an Obervational Approach

Because the prevailing views of explanations and of their properties run into fundamental difficulties when faced with the history-of-science explanation I am led to abandon not only those views, but also the theoretical approach to history-of-science explanation. That is, I shall abandon the attempt to apply to history-of-science explanation certain abstract and general principles of explanation that impose upon history-of-science explanation the structure of explanation in other fields. I shall instead begin not with theoretical (philosophical or methodological) principles but with data, the data being explanations actually put forth by historians of science. Moreover, since my investigation is normative as well as descriptive, I shall not only have to find my data but determine whether these data-explanations are satisfactory.

But how shall I select and evaluate my data? For the evaluation I do not have any explicit positive criteria, only the following two negative ones. First, I should avoid authoritarianism; that is, I should not regard an explanation as satisfactory simply because it is put forth by a well-known or otherwise influential historian of science; second, I should avoid methodological externalism by not regarding an explanation as unsatisfactory simply because it does not have the structure or characteristics that explanations in fields other than history of science have.

For the selection of the data I shall use the following principles. First in order of importance, the data must be recent or contemporary historical accounts which are put forth as explanations. Second, each account should be concerned with explaining an important type of historical event, fact, or phenomenon. Third, the thing being explained should have specific importance besides the general importance due to its type. And finally, the examples should be taken from the most reputable sources.

Three somewhat arbitrary but fairly representative explanations, selected on these principles, are: Henry Guerlac's *Lavoisier—The Crucial Year*,[5] Alexandre Koyré's second volume of *Études Galiléennes*,[6] and the explanation of the rise of modern science. I shall now determine whether these explanations are satisfactory.

chapter 5

Guerlac on Lavoisier

IN MY CRITICAL analysis of Guerlac's *Lavoisier—The Crucial Year*,[1] I shall try to extract from that book answers to the following questions: (1) What is Guerlac's aim? (2) What is his problem? (3) What is his explanatory thesis? (4) How satisfactory is the explanation? I shall argue that Guerlac's explanation is unacceptable on a number of specific but important points, and moreover that it is methodologically objectionable in several respects. I shall then state the results of my criticism in the form of an alternative explanation which accepts what is valuable in Guerlac's account and avoids his methodological errors. Finally, I shall sketch a more radical explanatory hypothesis which rejects even the more plausible elements of Guerlac's story and accepts only the relevant documents and texts from Lavoisier which Guerlac provided in an appendix to his book.

Guerlac's Aim

Guerlac describes his aim as that of explaining something, and of finding the solution to a historical puzzle:

> It has never been satisfactorily *explained* just how Lavoisier was led to carry out, in the autumn 1772, those first experiments on the burning of phosphorus and sulphur and on the reduction of the calx of lead which brought him in succeeding years to the discovery of the role of oxygen, to his antiphlogistic theory of combustion, and to a radical refashioning of the science of chemistry. I believe it is possible, despite the scarcity of documents from Lavoisier's own hand for the period in question, *to find a solution to this important historical puzzle.* [P. 1, my italics]

The above quotation is neither a chance remark nor an afterthought on Guerlac's part since other explicit statements recur throughout the book, expressing his explanatory intentions, summarizing the explanatory achievements up to a given point, and outlining the remaining explanatory task (pp. xiii, 76, 90, 146, 156, 192). Unquestionably his aim is to explain a certain event in the history of science.

The Problem

Guerlac's explicandum is Lavoisier's epoch-making experiments of the autumn of 1772, which Guerlac describes as follows:

> On November 2, 1772, as everyone familiar with the subject has long known, Lavoisier deposited with the Perpetual Secretary of the Academy of Sciences the famous sealed note (*pli cacheté*), dated the previous day, in which he briefly recorded his momentous discovery that when phosphorus and sulphur are burned they gain markedly in weight because of the "prodigious quantity of air that is fixed during combustion and combines with the vapours." This, he continued, "made me think that what was observed in the combustion of sulphur and phosorus could well occur in the case of all bodies that gain weight by combustion and calcination; and I became convinced that the increase in weight of metallic calces was due to the same cause." Lavoisier then described how he had confirmed this conjecture for the case of lead by reducing "litharge" (lead oxide) in a closed vessel, using the "apparatus of M. Hales," and had observed the considerable quantity of air given off at the instant the calx

changed into the metal. This discovery seemed to him "one of the most interesting made since Stahl." [P. 5]

Moreover, on the basis of the November 1 sealed note and other documents, Guerlac argues that the phosphorus experiments were performed sometime between September 10 and October 20 and "most probably after the middle of October" (p. 7), the sulphur experiment on or about October 24, and the calx-of-lead experiment between October 24 and November 1.

The genuineness of the problem, i.e., the really puzzling character of Lavoisier's cognitive behavior, is emphasized by Guerlac when he draws attention to the fact that "down to 1772—a year that Lavoisier himself took as pivotal in his career—he had displayed no interest in combustion or the calcination of metals and no curiosity about hints in the chemical literature that air might be worth the attention of a serious chemist" (pp. 3–4). On the contrary, Guerlac continues, Lavoisier had been primarily interested in geology as his first paper on gypsum and his contribution to the study of water indicate; not only had he displayed no interest in combustion and calcination but also as late as the spring of 1771 not even his plans for future research included those subjects. Instead they dealt with niter, indigo, barometric variation, improvement of hydrometers, and the lighting of cities.

The Explanatory Thesis

Guerlac's explanation involves two steps corresponding to the two primary aspects of the explicandum: Lavoisier's performance and interpretation of the litharge experiment, and the performance and interpretation of his experiments in burning sulphur and phosphorus. He first accounts for the litharge experiment and then explains (away) the combustion experiments.

The litharge experiment is accounted for by saying that Lavoisier performed it to test his new antiphlogistic hypothesis explaining the effervescence of calxes in reduction, the limited calcination of metals in closed vessels, and the calx augmentation effect. It is important to notice that Guerlac thus explains at once both Lavoisier's performance *and* his interpretation of the experiment. He explains the performance insofar as he is claiming that Lavoisier made the experiment in order to test something. And he accounts for Lavoisier's interpretation insofar as what he was trying to test was whether certain phenomena could be interpreted antiphlogistically. This is one of the most unsatisfactory ways of explaining scientific discoveries, as I shall show.

But why did Lavoisier devise his antiphlogistic hypothesis? Guerlac tells us the reason was that he found Guyton's phlogistic explanation of the augmentation effect highly unsatisfactory and even fanciful (pp. 111, 112, 194). And how did Lavoisier conceive of the antiphlogistic explanation? We are told that Lavoisier accomplished this by "combining" in his mind one idea and the knowledge of three facts: the idea that air plays an important role in chemical processes and the facts that metallic calxes effervesce during reduction, that metals resist calcination in closed vessels, and that the calx of a metal is heavier than the regulus. According to Guerlac there was essentially only one intermediate step in this "combining." Lavoisier is alleged to have first combined the idea of a chemical air with the knowledge of the fact that metallic calxes effervesce when reduced, to come to suspect that air is released in reduction and absorbed in calcination. Guerlac then accounts for Lavoisier's knowledge of those facts and conviction of the chemical role of air. He came to know about the augmentation effect and the limited calcination as a result of the discussions evoked in 1771/72 by Guyton's experimental results; the effervescence of calxes during reduction was common knowledge among chemists at the time, and Lavoisier either found it described in the chemical literature or observed it directly. The conviction about the chemical role of air, on the other hand, was not common; Lavoisier acquired it when the excitement about and interest in air and effervescence caused by the publication of Priestley's "Directions for Impregnating Water with Fixed Air" induced him to study directly the phenomenon of effervescence by investigating its occurrence in a number of common inorganic reactions. This was the form and extent of Priestley's influence on Lavoisier in the summer of 1772 and not at all like the influence on other French chemists such as Bayen, Bucquet, and the younger Rouelle, who were led to study "fixed air." Guerlac accounts ingeniously for the direction of this influence on Lavoisier by alleging that he had some direct or indirect acquaintance (but it was a merely vague acquaintance) with Hales's doctrine of fixed air of the *Vegetable Staticks,* which he probably consulted carefully after his interest in air and effervescence had been aroused by the "Directions."

The coherence of the account should not be overlooked; it ties together the influence of Hales, Priestley, and Guyton, the newly discovered document "System of the Elements" and the newly appreciated (by Guerlac) August memorandum, and the phenomena of effervescence, calcination, and augmentation. Finally, it provides the basis for the explanation of the combustion experiments. The crucial claims in this explanation involve the occurrence of two "accidents." The first is the unavailability of a sufficiently powerful burning glass until late October, which date Guerlac assumes is the earliest on which Lavoisier could put in practice his intention to carry out the litharge experiment envisaged in August. The second accident, one of Lavoisier's professional life, occurred while he was evaluating his friend Mitouard's work on phosphorus in order to sponsor him for membership in the Academy of Sciences. Mitouard's casual suggestion that the addition of air might explain the greater weight of phosphoric acid led Lavoisier to extend the scope of his own antiphlogistic explanation to include the burning of phosphorus. It was this possibility that he was investigating in the experiments on the burning of phosphorus.

Guerlac's explanatory account, which will serve as a basis for my criticism, can best be reconstructed in a temporally progressive narration and, for the most part, in his own words:

(L1) (The Hales-influence thesis)
". . . it was through Hales—indirectly through Boerhaave, Rouelle, Macquer, and others, if not directly through a reading of the *Vegetable Staticks*—that Lavoisier came to appreciate the important role that air might play in chemical processes. . . . But if Lavoisier, like those other French chemists—those who accepted Hales's views and those who did not—had long been familiar with the 'Analysis of Air', it is evident that something must have occurred in the months before Lavoisier wrote the August memorandum to give new meaning to the half-century-old experiments of the English parson." [Pp. 34–35]

(L2) (The Priestley-influence thesis)
"It was probably the first news of Priestley's discoveries, vague though they were, but more especially the appearance in France of Priestley's *Directions,* that aroused his interest in air, led him to turn to Hales's *Vegetable Staticks* for more information and induced him to perform those experiments to which he refers at the beginning of his 'Système sur les élémens'." [Pp. 100–101]

(L3) (The chemical-air-conviction thesis)
"Lavoisier's 'Système sur les élémens' makes it crystal clear that the phenomenon of effervescence, observed in a number of common inorganic reactions, had convinced him, by the early sum-

mer of 1772, that air must play a role in chemical processes. Yet there is no suggestion in this interesting document that air must be chemically involved in combustions and calcinations; indeed, such processes are not so much as mentioned. But by the time he set down the memorandum of August 8, 1772, this possibility [as regards calcination] had clearly occurred to him." [Pp. 101–102]

The following had occurred in the meantime (pp. 105, 111).

(L4) (The effervescence-in-reduction thesis)
Lavoisier became acquainted with the fact that effervescence-like phenomena can also be observed in the reduction of metallic calces. [Pp. 106–10, 194]

(L5) (The limited-calcination thesis)
Lavoisier also learned about "the fact (which Guyton had supported) that metals resist calcination in closed vessels." [P. 145, see also p. 135]

(L6) (The augmentation-effect thesis)
Most important of all, ". . . he had also become aware at about the same time of the significant fact that some metals, perhaps all metals that can be burned gain weight when they are calcined." [P. 111]

(L7) (The believed-incompatibility thesis)
Lavoisier immediately became aware too of "the evident incompatibility between the newly-established *fait capital* and the phlogiston theory to which Guyton tenaciously adhered." [P. 141]

(L8) (The contradiction-motivation thesis)
"Guyton's procrustean efforts to accommodate theory to experiment . . . caused Lavoisier to ponder the inherent contradiction and to search for a new explanation of the phenomenon itself." [P. 141]

(L9) (The antiphlogistic-explanation thesis)
"Sometime in the summer of 1772, before he wrote the August memorandum, several factors we have suggested—the effervescence of calxes during reduction, the fact (which Guyton had supported) that metals resist calcination in closed vessels, the first inklings of work on gases being carried abroad, and Hales doctrine of fixed air—combined to suggest to Lavoisier a more likely explanation: that the air fixed in metals during calcination and released upon reduction might account for the greater weight of a calx." [Pp. 144–45]

(L10) (The antiphlogistic-test thesis)

"As early as August 8 of 1772, he had envisaged an experiment to test this hypothesis." [P. 194]

(L11) "Yet it is well known, and we have repeatedly emphasized, that the first experiments he actually performed were those on phosphorus and sulphur and that the famous reduction of minium [calx of lead] was not carried out until late October." [P. 156]

(L12) (The forced-delay thesis)
"I believe that Lavoisier was obliged to postpone his projected experiment on lead oxide until he could have the exclusive use of the Academy's burning glass, with which he and his older collaborators had been experimenting during the summer and the early autumn." [P. 195]

(L13) (The extension-possibility thesis)
"The accident of having to evaluate Mitouard's work on phosphorus, perhaps stimulated by Mitouard's remark about the possible role of air [i.e., "Mitouard's casual suggestion that the addition of air might explain the greater weight of phosphoric acid" (p. 189)] gave Lavoisier the unexpected opportunity as well as the incentive to explore an interesting possibility: namely that what he believed to take place in the calcination and reduction of metals might also be true, as Mitouard seemed to think possible, in the case of burning phosphorus, a substance, incidentally, that could be readily studied without the use of the burning glass." [Pp. 189–90]

(L14) (The irrelevance-of-order thesis)
"It seems, therefore, to have been a mere accident that this hypothesis ["the hypothesis he proposed to test with the burning glass" (p. 191)] was first confirmed with phosphorus and then with sulphur, rather than by the experiment on the reduction of metallic calxes which he had envisaged as early as August 8, 1772." [P. 191]

Substantive Criticism

Without questioning for the moment the truth of Guerlac's account, I hold his explanation to be unsatisfactory because it is *significantly* incomplete. It says nothing about the sulphur experiment; that is, this aspect of the explicandum is left unexplained. I emphasize the word *significantly* because I think that no explanation can be expected to explain its explicandum in all of its aspects, the number of them which can be discerned in any one explicandum being, in general, indefinite, not to say infinite. For each aspect derives its reality from a possible contrast to which the explicandum may be subjected or under which it may be considered. For example, in our case, among the aspects that could be discerned are these:

(1) The occurrence (that is, the performance) of the experiment. The contrast here would be the nonoccurrence of those experiments; the question would be, why were those experiments performed at all?

(2) The way Lavoisier interprets those experiments, that is, the theoretical significance he attaches to them. The contrast here would be the performance of those experiments or observations without the interpretation or theoretical significance Lavoisier attached to them; the question would then be, why did Lavoisier interpret the experiments or observations he made in the way he did? (1) and (2) are the aspects on which Guerlac seems to be concentrating.

(3) The temporal order of the experiments. Several sub-aspects are included under this heading, each one corresponding to a different possible order in which the experiments might have been performed, e.g., calx of lead first, phosphorus second, sulphur third or calx of lead first, sulphur second, phosphorus third, etc.; the question would be, why did Lavoisier perform the experiments in the order O in which he did rather than in some other order O'?

(4) The timing of the experiments. This aspect is related to the previous one but is distinct from it. The contrast here would be Lavoisier's performance of those experiments at some other time, say the summer of 1772 or the year 1771. The question would be, why did Lavoisier perform his epoch-making experiments in the autumn of 1772 rather than in the summer? The "delay" in the reduction experiment, which Guerlac explicitly recognizes as an aspect of the explicandum worth explaining, would be a sub-aspect of this general aspect.

(5) The personality or the individuality of the experimenter. Here the contrast would be some other scientist than Lavoisier; the question would be, why did Lavoisier rather than someone else, say Turgot, or Guyton, or Priestley, perform and interpret those experiments in the way he did? Guerlac seems to think that he has explained this aspect of the explicandum when he says, "I have tried to identify the accepted beliefs and prejudices which prevented others from thinking as he came to think" (p. 192). But I think that this remark is a self-misrepresentation for I do not see that he has done anything of the sort.

(6) The choices of substances. The contrast here would be the choice of some other set of substances, such as calx of lead only, calx of lead and phosphorus only, calx of lead and sulphur only, calx of lead and lead metal, calx of lead and calx of mercury, phosphorus and sulphur

only, etc. One question would be, for example, why did Lavoisier perform the set of experiments he did rather than the reduction only, that is, why did he perform the combustion experiments at all? Another question would be, why did Lavoisier perform the experiments he did rather than the reduction and the phosphorus experiments only? That is, why did he perform the sulphur experiment at all? And it should be noted that, whereas Guerlac tries to answer the former question, he does *not* answer the latter.

The real question then is, how important is the sulphur aspect of the explicandum? Guerlac must claim that it is not important unless he wants to admit that his explanation is significantly incomplete. I am claiming that it is important. Who is right? If it were only a question of judgment, then Guerlac's judgment should certainly prevail over mine. But it is not just a question of judgment; it is also a question of argument. I would argue as follows. First, Guerlac himself obviously believes that the phosphorus aspect is important and thus worth explaining. But if the phosphorus aspect is worth explaining why is not the sulphur aspect also worth explaining since both aspects are sub-aspects of either the choice of substance or of the combustion aspect of the explicandum? Guerlac might reply that Lavoisier himself thought that the phosphorus experiment was more important as shown by the fact that both the September 10 and the October 20 notes concern phosphorus but do not mention sulphur and, moreover, that the October one is rather long, a "memoir-torso," as the scholar Speter called it. I would answer that the sealed note of November 1 describes the sulphur experiment in more detail than the phosphorus one, which is only casually mentioned. And certainly the sealed note must be regarded as more important and as more representative of Lavoisier's judgment. Therefore, the evidence from Lavoisier indicates at worst (for me) that phosphorus and sulphur are of comparable importance but quite possibly that the sulphur aspect is more important than the phosphorus aspect.

Second, Lavoisier himself says in the November note that he was "led to think that what is observed in the combustion of sulphur and phosphorus may well take place in the case of all substances that gain weight by combustion and calcination." But the difference in soundness as well as in psychological ability between generalizing from two cases and generalizing from one case is in fact quite remarkable, much more than some "inductive logician" would be led to believe from his formal calculus. Therefore, it is important that Lavoisier performed two combustion experiments and consequently the sulphur experiment is a non-negligible part of the explicandum.

The truth of the matter is that the question of importance is not altogether independent of the available explanation. For example, it may be that one cannot be sure of the abstract and objective importance of a certain aspect of an explicandum if one has no explanation of the explicandum as a whole; or it may be that one cannot be sure of such importance if one has no explanation of that same aspect. Conversely, if one has an explanation of certain aspects of an explicandum, one may be inclined, and perhaps rightly in some cases, to regard only these explained aspects as important and the unexplained ones as unimportant. The latter is Guerlac's case, for he himself says (though in connection with the relative importance of the phosphorus and of the litharge aspects), "If Lavoisier, in the first instance, planned to confirm his suspicion by an experiment on the reduction of [calx of] lead—and I have marshalled considerable evidence to show that he did—then the question of Lavoisier's interest in phosphorus, and the whole Mitouard episode, are both diminished in significance" (p. 196). Also diminished in significance, but still in need of an explanation would be Lavoisier's performance of the sulphur experiment. This Guerlac has not provided. Therefore, in any case, Guerlac's explanation remains incomplete.

It is incomplete in another important respect as well. I think that it is misleadingly incorrect for Guerlac to claim in the believed-incompatibility thesis (L7) that sometime between the writing of the "System of the Elements" (end of July) and the writing of the August 8 memorandum Lavoisier became aware of "the evident incompatibility between the newly-established *fait capital* and the phlogiston theory to which Guyton tenaciously adhered" (p. 141). The thesis should be phrased to state that "Lavoisier became convinced of the incompatibility between . . ."; this is all that Guerlac need claim at this particular place in his account. Not that he does not also hold this revised thesis; he does. But for him it does not much matter which statement he makes since he seems to be of the opinion that the augmentation effect is evidently incompatible with the phlogiston theory: "Here was a phenomenon which the phlogiston theory could not adequately explain; for according to this widely held theory, burning and calcination were accompanied by a *loss* of this hypothetical fire principle, whereas an increase in weight should prove that something has been added" (p. 111, see also pp. 141, 194).

But the augmentation effect is neither "evidently" nor *at all* incompatible with (Guyton's) phlogiston theory. It is not "evidently" incompatible since the notion of a negative weight as an abstraction is by no means chimerical, unless one wants to say that the modern physical notion of energy, for example, is chimerical. (Consider the case of an artificial satellite which by losing energy becomes heavier and eventually falls to the ground.) That it is not *at all* incompatible is shown by the fact that the loss of hydrogen or helium by a balloon makes it heavier because of Archimedes' principle and the smaller-than-air specific gravity of hydrogen or helium. This was in fact how many of the phlogistonists argued.

Since the augmentation effect is not "evidently" incompatible with (Guyton's) phlogiston theory, it is crucial to distinguish between Lavoisier's becoming aware and his becoming convinced of the said incompatibility, and to say which we mean. Therefore, as stated, Guerlac's believed-incompatibility thesis is incorrect since one cannot become aware of what is not so; one can only, mistakenly perhaps, convince oneself of what is not so. However, since elsewhere in the book Guerlac makes only the weaker claim that Lavoisier became convinced of the incompatibility, we can take this to be the meaning of Guerlac's thesis and regard only as misleading and not actually incorrect the passages where he makes the stronger claim.

We have exhibited above Guerlac's error or his superficiality in his belief in the incompatibility of the augmentation effect and the phlogiston theory. This error turns out to be irrelevant to the truth or correctness of Guerlac's explanation since the believed-incompatibility thesis (L7) which is intended to be part of the explanation turns out to be only misleadingly phrased and could easily be correctly rephrased, in accordance with Guerlac's meaning and intention. But when correctly reformulated, (L7) requires explanation. This, however, Guerlac nowhere notes in his book; of course, he never accounts for Lavoisier's error of believing in the incompatibility because, in view of his own sharing of the error, there is nothing to account for. Thus Guerlac's error, though essentially irrelevant to the truth, is highly relevant to the completeness and/or comprehensibility of his explanation. His account actually fails to make the explicandum intelligible to the extent that the fact alleged by (L7) is not made easily understandable. Although explicable in terms of Guerlac's error, this incompleteness not only remains unjustifiable but, when added to the one previously exhibited, begins to weigh and affects the evaluation of

Guerlac's explanation. I shall, however, say nothing further about this aspect of the explanation since its falsity, which will shortly become evident, gives such criticism only secondary interest.

Many of the theses, including all of the crucial ones constituting Guerlac's explanation, are, in fact, irrevocably inaccurate. More specifically, the effervescence-in-reduction (L4) thesis is unfounded, whereas the limited-calcination thesis (L5), the augmentation-effect (L6), the antiphlogistic-explanation (L9), and the antiphlogistic-test (L10) theses are very probably false; finally, the inaccuracies in Guerlac's explanation of the litharge experiment spread and infect his account of the combustion experiments.

I shall examine first the effervescence-in-reduction thesis (L4). Let me emphasize here that what is at issue is whether, in early August 1772, Lavoisier knew that the phenomenon of effervescence can be observed in the reduction of metallic calxes. In its support Guerlac gives an argument (pp. 105–106) based on the evidence of certain statements made by Lavoisier in *Opuscules physiques et chimiques* of 1774! That the evidence cannot be trusted and that Guerlac's argument cannot be taken seriously is shown by Guerlac's own claim that in the *Opuscules* Lavoisier confuses the historical with the logical order of events. As Guerlac notes, whereas in the *Opuscules* Lavoisier suggests that the experiments on the increased weight of metallic precipitates from solution were performed before his speculations on the augmentation effect (1772), Lavoisier's *Registres* make it clear that they date from the spring and summer of 1773 (p. 106 n.).

This is one of the many instances of Guerlac's inconsistent and opportunistic attitude toward the evidence of Lavoisier's own words. Sometimes he is skeptical about what Lavoisier says he does, did, or has done; for example, Guerlac thinks that Lavoisier's own account of his train of thought as stated in the sealed note of November 1 is untrustworthy and misleading. Guerlac's whole book rests on that assumption. But although he rejects evidence of what Lavoisier *says* in this case where what he says is very intimately connected with, and hardly distinguishable from, what he *does*, Guerlac accepts the evidence of Lavoisier's words in other cases where it suits his purposes, although in such cases what Lavoisier says is readily distinguishable from what he does. One such case, besides the one noted above, is Lavoisier's own words in the "Reflections on the Phlogiston" of 1783, used by Guerlac to support his antiphlogistic-explanation thesis (L9) (pp. 104–105). I conclude that Guerlac has provided us with no good reasons to accept (L4).

I turn now to the limited-calcination thesis (L5). The limited calcination of metals in closed vessels *may* be explained by Lavoisier's antiphlogistic hypothesis (which it is), and it may have been established as a fact in early August 1772 (which I think it was not); but the crucial claim here is that between the writing of the "System of the Elements" (late July according to Guerlac's dating) and the writing of the August 8 memorandum, Lavoisier came to accept limited calcination as a fact-to-be-explained. I do not wish to magnify the importance Guerlac attributes to (L5), but limited-calcination is one of the three facts which together with an idea (the idea of the chemical role of air) "combined" in Lavoisier's mind to yield the antiphlogistic-explanation hypothesis (L9) that he was testing by the litharge experiment. It is regrettable therefore that Guerlac *does not explicitly* (even badly) argue in its support. This betrays, I think, a serious (fatal, as we shall see) confusion between logical and historical questions. That is, Guerlac seems to be confusing the logical fact that limited calcination is one of the phenomena explicable by the antiphlogistic hypothesis with the historical thesis that it was one of the facts known to Lavoisier which led him to his hypothesis. More of this below.

Be that as it may, Guerlac hints (pp. 102, 135) that the August memorandum is the evidence on which he is basing his claim. If we examine that document, we discover that references to calcination occur only in two passages in that document (and note that, as I shall emphasize later, the last section of that document is *not* one of these places as Guerlac thinks). The first passage is from the introductory section:

> The fire which chemists usually employ can neither be lighted nor subsist in a vacuum. Air is a necessary agent for its preservation. The fire of the burning glass offers in this respect a very great advantage. It can penetrate under the receiver of the pneumatic machine, and by means of it one can make calcinations and combinations in a vacuum. [P. 210]

The second passage is the section entitled "On Metals."

> Independently of M. Homberg's and M. Geoffroy's experiments with metals, which it will be necessary to repeat, it will be a good

thing to check whether they can be calcined in closed vessels. They all give off a vapor or fume when placed at the focus of a burning glass; it would be very interesting to find an apparatus appropriate to retain and condense it. Vessels made out of rock crystals might hold the object, but one must make sure in advance that they withstand the effects of the burning glass. [P. 212]

Each passage *refutes* (L5). For, the first passage, insofar as it is relevant, suggests that Lavoisier thinks that calcinations are possible in a vacuum, let alone in a limited amount of air. What seems to worry Lavoisier about the lack of air is not the limitations on calcination, but the unattainability of the ordinary source of heat, fire. But the heat of the burning glass needs no air, and should, therefore, allow one to make calcinations in its absence.

Insofar as the second passage is relevant, and in Lavoisier's mind it clearly is no more relevant than the first, Lavoisier is not sure whether the metals can be calcined in closed vessels. Insofar as he is more inclined one way than the other, he expects that they can, for he expects to collect and condense the vapors and fumes given off during these calcinations. His expectation, incidentally, shows where his real interest lay: the constitution of matter and air as a constituent thereof. The limited-calcination thesis (L5) is, therefore, false: Lavoisier had learned and was aware of no fact concerning the resistance of metals to calcination in closed vessels.

It is easy, I think, to overestimate the extent to which the augmentation-effect thesis (L6) is well supported. Guerlac himself admits (pp. 142–44) that there is practically no evidence for it. That the augmentation effect was "of crucial importance in the crystallization of his thought" (p. 111) no one doubts. But the crucial claim that Guerlac needs for his explanation is that Lavoisier became aware of the augmentation effect between the writing of the "System" and of the August 8 memorandum. Therefore, it is irrelevant for Guerlac to point out that Lavoisier "mentioned it in the sealed note of November; and in the *Opuscules* of 1774 he singled it out as of real importance" (p. 111). The passage from the "Reflections on the Phlogiston" of 1783, quoted by Guerlac on page 112, is worse than irrelevant, for it shows that by that time the augmentation effect had "become an incontestable truth." The fact that in a paper on the precipitation of metals read to the Academy in 1783, Lavoisier makes use of Guyton's figures for the characteristic increase in the weight of different metals (p. 112) only shows that, after Lavoisier's antiphlogistic hypothesis had become established, he could reinterpret Guyton's original phlogistically interpreted results to support his own theory. The same conclusion is to be drawn from Lavoisier's sending Guyton a copy of the *Opuscules* complimenting him, alas, for his "observational genius" (p. 143).

The evidence of the *undated* note "On the matter in fire" remains. What the note conclusively shows is that Lavoisier must have become aware of the augmentation effect before he conceived of its explanation, i.e., before November 1; for in it Lavoisier says, referring to the augmentation effect, "Whatever its explanation is the fact is no less regular" (quoted by Guerlac, p. 144). It is not sufficient for Guerlac that the undated note should "have been written at roughly the same time as the August memorandum," as he "guesses" on page 144. In fact he modifies his "guess" to conclude in the next paragraph that "sometime in the summer of 1772 *before* he wrote the August memorandum several factors . . . combined" (p. 144, my italics).

At any rate (L6) can be independently disconfirmed by the fact that in the October 20 draft memoir Lavoisier terms the increase in weight of phosphorus when burned a "singular" phenomenon (p. 225). Obviously he is not aware of any weight augmentation in the calcination of metals. I think, therefore, that the thesis is less than unfounded, it is false.

The unfoundedness of (L4) and the falsity of both (L6) and (L5) spread and infect the antiphlogistic-explanation thesis. For, obviously, no facts or factors can "combine" with each other in someone's mind unless he is actually thinking of them. But these faults of (L9), though serious enough, are the least of its faults. It is, in fact, utterly false in its central claim that before the writing of the August memorandum Lavoisier conceived of explaining the augmentation effect antiphlogistically. This can be seen by an examination of the document the interpretation or misinterpretation of which constitutes Guerlac's argument in support of the thesis. But first we shall clarify the exact role of this argument.

As I have noted previously, Guerlac alleges that one idea, the notion of the chemical role of the air, and one fact, the effervescence of metallic calces in reduction, combined first in Lavoisier's mind so that he "came to suspect, in the summer of 1772, that air is given off when metallic calxes are reduced, and consequently that it is absorbed when metals are burned or roasted" (p. 111, see also pp. 104 and 194). I shall call this the air-in-reduction-calcination thesis. As it happens, this is the only part of the antiphlogistic-explanation thesis (L9) that Guerlac explicitly argues for. This could be taken as evidence that Guerlac is confusing Lavoisier's conception of the antiphlogistic explanation with the above mentioned suspicion concerning the role of air in calcination and reduction. But I shall not emphasize this relatively minor criticism. The main point at issue here is whether the air-in-reduction-calcination thesis provides support for the central claim of (L9).

It seems clear that support for (L9) would provide support for the air-in-reduction-calcination thesis, since the latter is a rather obvious consequence of the former. But to support the latter is not thereby to support the former since (L9) is not a rather obvious consequence of the air-in-reduction-calcination thesis. In other words, the logical relation between the beliefs attributed to Lavoisier by the two theses is such that, though to show that he held the antiphlogistic explanation might be taken to support that he held the air-in-reduction-calcination hypothesis, to show that he had conceived of the air-in-reduction-calcination hypothesis is definitely not to support the claim that he had conceived of the antiphlogistic explanation.

Nor can Guerlac be rescued here by saying that, though to provide support for the air-in-reduction-calcination thesis is *not* thereby to provide support for (L9), to have provided support for the air-in-reduction-calcination *and* for the augmentation effect (L6) *and* for the limited-calcination (L5) theses *is* to support (L9); for, as we have seen, both (L5) and (L6) are probably false.

The above criticism is meant to expose the formal-logical fallacy of Guerlac's argument, perhaps a trivial matter, though, in this case in which we are dealing with a historical agent's thinking, it could be argued to be not so trivial. The criticism, however, is meant to expose the historical fallacy in Guerlac's argument as well. For we can expect, as Guerlac should have been watchful for, that Lavoisier lacked information and had "prejudices" which prevented his equating the air-in-calcination-reduction hypothesis with the antiphlogistic explanation. We have seen already that he probably lacked knowledge of the augmentation effect in the first place; it is also clear that he still had some phlogistic "prejudices" as can be seen in the October memoir-torso on phosphorus, but more importantly from the beginning sections of the August memorandum itself. Finally, it is clear from the passage entitled "On Metals," quoted above from the August memorandum, that Lavoisier believed that metals are somewhat volatile, that they fume and vaporize when strongly heated. Guerlac himself uses these volatility "prejudices" of Lavoisier's to refute his own earlier account of Lavoisier's investigations on the disappearance of diamond when strongly heated (pp. 87–89). But it is clear, then, that if Lavoisier believed that metals (perhaps partially) volatilize when heated he could not even begin to account for the augmentation effect even if he should have also believed that they absorb air when they are burned.

But I have not yet examined whether Guerlac has even shown this much, namely his air-in-calcination-reduction thesis, which is a subthesis of the antiphlogistic-explanation thesis (L9), and which attributes to Lavoisier an hypothesis alleged by Guerlac to be an intermediate step in Lavoisier's "combination" of the three facts and one idea. The first thing to notice about the air-in-reduction-calcination thesis is that it attributes to Lavoisier not only the suspicion "that air is given off when metallic calxes are reduced" (p. 111), which I shall call the air-release-in-reduction subthesis, but also the suspicion "consequently that it is absorbed when metals are burned or roasted" (p. 111), which I shall call the air-absorption-in-calcination subthesis. The second point to notice is that the only real argument that Guerlac gives in support of the *thesis* is one based on his interpretation (misinterpretation, as we shall see) of the last section of the August memorandum (p. 104). But this argument at best supports only the air-absorption-in-calcination subthesis, "that Lavoisier had come to suspect . . . that metals might absorb air in being transformed into their calxes" (p. 104). To be sure, Guerlac gives, also in support of this subthesis, another attempted argument based on the evidence of Lavoisier's allegations in the "Reflections on the Phlogiston" of 1783 (pp. 104–105). But besides being methodologically wrong and opportunistic in using this evidence Guerlac fails to see that the evidence, *if* methodologically acceptable, would support only the air-release-in-reduction subthesis. Here is the passage quoted by Guerlac:

> Something which constantly occurs in all metallic reductions led me to make some investigations on this matter: I noticed that, in all these processes, there was considerable effervescence at the moment when the metal passes from the state of calx to the metallic state; it was natural to conclude from it that a gas was being given off, and I conceived an apparatus appropriate to collect it. [P. 105]

The root of the difficulty seems to be once again, that Guerlac has confused logical with historical issues. Does he give evidence to convince us that, in the summer of 1772, Lavoisier would have been able to infer the absorption of air during the calcination of metals from the release of air during the reduction of calxes, and conversely? To be historically minded one would have to be on the alert and expect Lavoisier to have been at the time unable to see what we can see, or what he himself came to see (in November, say), namely that the heating of calxes and heating of metals involve opposite processes. Why should they be opposite processes according to Lavoisier? After all they are both heating processes. This is precisely the way Lavoisier thinks of them as shown in the August memorandum where he is interested in applying the "superior fire" of the burning glass to the various metallic and mineral substances, rocks, earths, minerals, in order to determine their constitution, which is what he is really after as we know also from the "System of the Elements" (late July). This can be seen from the way the August memorandum starts. Lavoisier first notes that Stahl's "famous and worthy of so being" phlogiston theory might seem more acceptable if one compared it to the ideas of the Frenchman M. Geoffroy senior whose "observation . . . had led him to conclude that all metals consist . . . (1) of a vitrifiable earth, specific to each, (2) of an oil or flammable principle" (p. 209). But, Lavoisier continues:

> Whether this system was that of Stahl, or whether M. Geoffroy was its inventor, it remains true that the experiments made with the burning glass lead one to it; and that suffices to emphasize how important this kind of experiment is.
> The action of this fire, superior to that which we employ in our laboratories, has so far been applied only to metallic substances; no systematic experiments at all have been made with rocks, ores, and infinitely many mineral substances. Moreover, the few experiments which are known to us have been made in open air without introducing any variety in the operations. The experiments with the burning glass offer then a still completely new course to follow. One will become more and more convinced of this from the following reflections. [Pp. 209–10]

And yet Guerlac is forced to claim that Lavoisier was able to make the above mentioned inference. The air-release-in-reduction subthesis, then, supported by Guerlac in a methodologically objectionable manner, is thus shown to be unsupported by the air-absorption-in-calcination subthesis. Nor can the air-release-in-reduction subthesis be independently supported on the basis of the chemical-air-conviction (L3) and effervescence-in-reduction (L4) theses for we have seen that (L4) is itself unfounded. Moreover, (L3) would be too weak and would have to be made more specific in order to yield, together with (L4), the air-release-in-reduction subthesis, which is therefore not only unsupported but especially in view of our evidence and argument above, probably unsupportable because falsified.

Guerlac's only remaining support, then, for the antiphlogistic-explanation thesis (L9) is the presumed truth of the air-absorption-in-calcination subthesis. Even if this subthesis were true it could not, of course, establish (L9), which claims much more, and which we have already found highly unlikely on other grounds—Lavoisier's volatility "prejudice" and his opinion of the "infamous" phlogiston theory as "famous and worthy of so being." In fact, the air-absorption-in-calcination subthesis, which is the only part of (L9), in support of which Guerlac argues, is none other than what he *means* by his chemical-air-conviction thesis (L3). True, this latter thesis claims that in the early summer 1772 Lavoisier had acquired the conviction "that air must play a role in chemical processes" (p. 101) and is correct insofar as it states that. But what Guerlac *means* and what he actually needs and uses in his explanation is only that "Lavoisier had come to suspect . . . that metals might *absorb* air in being transformed into their calxes" (p. 104), i.e., the air-absorption-in-calcination subthesis. This subthesis is Guerlac's worst blunder. I have already mentioned that it rests on his interpretation of the last section of the August memorandum. Lavoisier says: [1]

> It seems invariable that air is included in very great abundance in the constitution of most minerals and even of most metals. Yet no chemist has yet included air in the definition of either metals or any mineral bodies. An effervescence is nothing but a sudden release of the air which is in some way dissolved in each of the bodies that one combines.
>
> This release occurs whenever less air is contained in the constitution of a new compound than there was in each of the bodies which combine with each other. These views if pursued and elaborated could lead to an interesting theory which has been even already sketched; but here one must pay attention to the fact that most metals no longer effervesce after they have been kept for a long time at the fire of a burning mirror. Undoubtedly the degree of heat there experienced removes from them the air which is contained in their constitution. What is very odd is that metals in this state are no longer malleable and

they are virtually insoluble in acids. This observation which still needs confirmation can provide ample subject for observations and reflections.

It would be desirable that one be able to apply to the burning glass Hales's apparatus in order to measure the quantity of air produced or absorbed in each operation, but one fears that the difficulties facing this kind of experiment are insurmountable at the burning glass. [Pp. 103 and 214]

Guerlac interprets this passage as follows:

There are several things in this text worthy of attention besides the reference to the phenomenon of effervescence in connection with the behavior of metals. It is apparent that Lavoisier's earlier study of effervescences had made it seem likely that metals contain air, which can be released by treatment with acids. But what happens when a metal is transformed into its calx by exposure to sunlight in the focus of a burning glass? According to Stahl, it must simply lose phlogiston. Lavoisier suggests, however, that the heat may drive off the combined air, for this might explain the loss of malleability and of solubility in acids. Clearly, he is uncertain as to what takes place: he is only clear that air must be involved in some way in this transformation of a metal into its calx. Was it possible, as the last paragraph seems to imply, that air might be *absorbed*, rather than produced, during this calcination? To answer this question he proposes to subject metals and their calxes in a closed vessel to the heat of a burning glass, using some form of Hales's pedestal apparatus.

Is there stronger evidence than the August memorandum that Lavoisier had come to suspect, in the face of the observation that treating metals with acids produces air, that metals might *absorb* air in being transformed into their calxes? [P. 104]

The question "But what happens when a metal is transformed into its calx by exposure to sunlight in the focus of a burning glass?" is misleading insofar as it gives the impression that the subject of calcination is introduced directly by Lavoisier. On the contrary, Lavoisier introduces the subject of calcination into the discussion of air being a constituent of metals only indirectly, if at all, by reporting that, when placed in acid solutions, metals no longer effervesce if they have been "kept for a long time at the focus of a burning glass." There is no indication that Lavoisier is even aware that a metal "kept for a long time at the fire of a burning glass" is the calx of that metal. On the contrary, Lavoisier speaks of the *metal* being in a new state, for he says that "*metals in this state* are no longer malleable" (my italics).

In short, neither the word nor probably the concept of calcination is mentioned by Lavoisier in this passage.

The fifth sentence, "Lavoisier suggests, however, that the heat may drive off the combined air, for this might explain the loss of malleability and of solubility in acids," misrepresents the facts; it attributes to Lavoisier an argument which he neither gives nor would be likely to give. Guerlac portrays Lavoisier as inferring that the heat of the burning glass drives off the air combined with the metal on the grounds that this lack of air in the metal would explain its loss of malleability and of solubility in acids. That Lavoisier does not give such an argument is evident from the text. That it is unlikely that he would can be seen from the lack of indication that he would be prepared to make the explanatory claim that Guerlac attributes to him and on which Lavoisier would allegedly be grounding his inference. In fact, Lavoisier merely mentions the two phenomena, effervescence and the loss of malleability and solubility, without asserting a causal relation between the two. Again, it is unlikely that Lavoisier would make the explanatory claim since he is not certain that there is anything to explain, that is, he is not even absolutely sure of the reality of the loss of malleability and solubility phenomena. For the observation that "metals in this state are no longer malleable and they are virtually insoluble in acids" is the most likely reference for the word "observation" which is the subject of the sentence following it, "this observation which is still in need of confirmation. . . ."

Moreover the fifth statement is misrepresentative in the same way as the sixth. For Lavoisier's attitude is portrayed by Guerlac as "suggestive" in the fifth statement and as "uncertain" in the sixth, whereas the evidence of the text, "Undoubtedly [*sans doute*] the degree of heat there experienced removes from them the air which is contained in their constitution," seems to be that Lavoisier has no doubts on *that* matter. Even with a weaker interpretation of "sans doute" Lavoisier would still be reasonably certain, certain enough to be willing to say "probably." Even clearer is the evidence of the "System of the Elements," whose theory is actually the basis of Lavoisier's confidence.

The sixth statement is also inaccurate in suggesting in its second part, as the third sentence does, that Lavoisier is primarily or directly discussing the transformation of a metal into its calx, whereas what he is primarily discussing is the constitution of bodies, and of metals in particular, and air as a constituent thereof. Lavoisier is primarily interested in the constitution or structure of matter rather than in certain processes, though the two are related and Lavoisier is aware of it. In other words, insofar as Lavoisier is uncertain about anything, it is not what Guerlac alleges him to be uncertain about, namely, "what happens in calcination," but rather "whether heat drives off from metals their constituent air."

Having misinterpreted Lavoisier's attitude, Guerlac portrays him in the seventh statement as posing a question, "Was it possible . . . that air might be absorbed, rather than produced, during this calcination?" But Lavoisier poses no questions either explicitly or implicitly. In fact, as we have seen, he has few doubts, contrary to what Guerlac thinks.

And having misinterpreted the nature of Lavoisier's beliefs, Guerlac portrays him as being concerned with a different problem. Guerlac would have Lavoisier be concerned with the problem of determining whether air is produced or absorbed during the calcination of metals. This is inaccurate in two ways. First, Lavoisier is concerned with the problem of "applying Hales' apparatus to measure the quantity of air . . ." and not with "determining whether air. . . ." Second, Lavoisier is concerned with "air produced or absorbed in *each* operation (dans *chaque* operation)" and not with "air produced or absorbed in *this* operation," presumably calcination according to Guerlac. In other words, Lavoisier is concerned with applying Hales's apparatus to measure the quantity of air produced or absorbed in each operation. The question naturally arises, "In which operations?" The answer is provided by the title of the memorandum: "In 'the experiments [on various substances] which one can try with the help of the burning mirror.'" Which substances? Substances, like metals, which, according to Lavoisier's theory, produce air and substances, like calces perhaps, which, according to the theory, absorb air. To summarize: Lavoisier is concerned with applying Hales's apparatus to measure the quantity of air produced or absorbed in each operation depending on whether the operation is air-producing or air-absorbing, respectively. And this should not be surprising; for, if Lavoisier's discussion in that last section of the memorandum is "on the air contained in bodies," as the subtitle of that last section assures us, then if he wants to measure the amount of this air he must measure the air produced or absorbed in the appropriate reactions.

That Lavoisier's concern is with a measurement problem and not with the question of whether air is produced or absorbed during calcination is demonstrated by the fact that if, as we can agree with Guerlac, Lavoisier is "clear that air must be involved in some way in this transformation of a metal into its calx" (p. 104), and if, as Lavoisier is convinced, "an effervescence is nothing but a sudden release of the air which is in some way dissolved in each of the bodies that one combines" (pp. 103 and 214), then Lavoisier would not have needed Hales's apparatus to answer the question attributed to him by Guerlac for he could have simply *observed* whether the heated substance effervesces or not. If it does, then by Lavoisier's definition of effervescence, air is produced. If it does not, then air is being absorbed, it being the only remaining alternative according to Lavoisier's theory. Not needing Hales's apparatus to solve the problem Guerlac attributes to him, how could Lavoisier propose to use it for that purpose? How could he, especially when he suspects that the use of Hales's apparatus presents insurmountable problems, that is, when he "fears that the difficulties facing this kind of experiment are insurmountable at the burning glass?" But he would need Hales's apparatus if he had intended to measure the quantity of air produced when it is produced or absorbed when it is absorbed.

Finally, having misinterpreted the nature of Lavoisier's problem, Guerlac attributes to him the wrong purpose. For in the next to the last sentence of his interpretation quoted above, Guerlac says that "to answer this question [the question "Was it possible . . . that the air might be *absorbed*, rather than produced, during this calcination" (p. 104)] Lavoisier proposes to subject metals and their calxes in a closed vessel to the heat of a burning glass, using some form of Hales' pedestal apparatus" (p. 104). But, although Lavoisier does propose "to subject metals and their calxes in a closed vessel to the heat of a burning glass, using some form of Hales' pedestal apparatus," his purpose is, as we have seen, to measure the quantity of air produced or absorbed in each reaction of the corresponding type and not to answer the question of whether air might be absorbed, rather than produced, during calcination.

I conclude that Guerlac is mistaken concerning Lavoisier's interest, attitude, beliefs, and purpose, as expressed in the last section of the August memorandum.

It is clear what the consequences of this misinterpretation are for the air-absorption-in-calcination subthesis, which is the last, though feeble, support for the antiphlogistic-explanation thesis (L9), and which is the meaning if not the letter of the chemical-air-conviction thesis (L3). The opposite of the air-absorption-in-calcination subthesis, namely that Lavoisier believed on August 8, 1772, that metals release air when strongly heated, is what the August memorandum shows. Therefore, not only have we refuted Guerlac's argument in the support of the all-important (L9), but in so doing we have shown this thesis to be false on the grounds, to repeat, of Lavoisier's above-mentioned belief, of his volatility "prejudice," and his opinion of the phlogistic theory as "famous and worthy of so being."

The consequences for the antiphlogistic-test thesis (L10) are equally clear and immediate. Lavoisier was not envisaging on August 8, 1772, an experiment to test his antiphlogistic explanation. He was proposing an experiment, but an experiment intended to measure certain quantities, and such an experiment is not a "test." Nor could have Lavoisier envisaged a test of the antiphlogistic hypothesis because, as we have pointed out above, he had almost certainly not conceived of this hypothesis. Finally, neither could he have conceived of a test to answer "whether air was produced or absorbed in calcination," a question he was not asking since he thought he knew the answer, which we now know to be a mistaken answer. (L10) thus collapses.

It remains to be shown how the crumbling of Guerlac's account of the litharge experiment affects the explanation of the phosphorus experiments. I think there is no question that, though the Mitouard story may be unaffected, nothing of explanatory value is left in that story. The extension-possibility thesis (L13) falls to pieces since its principal claim is that Lavoisier performed the phosphorus experiments in order to explore the possibility that "what he believed to take place in the calcination and reduction of metals might also be true, as Mitouard seemed to think possible, in the case of burning phosphorus" (p. 190; see also p. 63 above). We have seen, in fact, that there is no reason to accept Guerlac's account of what Lavoisier believed to take place in calcination and reduction; he could not, therefore, have conjecturally generalized that hypothesis to the case of the burning of phosphorus. Similarly to be rejected is Guerlac's irrelevance-of-order thesis (L14) that it was an accident that Lavoisier's antiphlogistic hypothesis "was first confirmed with phosphorus and then with sulphur, rather than by the experiment on the reduction of metallic calxes which he had envisaged as early as August 8, 1772" (p. 191), for we have seen that Lavoisier had no such hypothesis to confirm. This is also clear from an examination of the October 20 draft memoir, which is still written in phlogistic terms, showing that, though he was making observations which later turned out to be significant, he hardly knew at the time "what he was doing" (seen from his later point of view). No wonder, then, that the phosphorus draft memoir remained a memoir-torso. No wonder, too, that the "System of the Elements" was never published and was even "lost."

Methodological Criticism

I have shown above the incompleteness and the fundamental incorrectness of Guerlac's explanation. It remains to examine to what extent these qualities of his account are due to methodological errors. I have already pointed out Guerlac's opportunism with respect to the evidence of Lavoisier's own words. To be sure this is only opportunism from the point of what I would regard as one of the supreme maxims of the history and the philosophy of science attributed to Einstein by S. Toulmin (*The Philosophy of Science*, p. 16): "If you want to find out anything from the theoretical physicists about the methods they use, I advise you to stick closely to one principle: don't listen to their words, fix your attention on their deeds." Perhaps another maxim could be formulated which would make consistent Guerlac's procedure and could confrom with it. But this is Guerlac's problem. I do not think that he could do it.

At any rate, opportunistic or not, I think Guerlac proceeds wrongly in all those cases where he accepts as evidence Lavoisier's own later words and "memories." On the other hand, his methodological opportunism and errors should not blind us to the real merit of his fundamental attitude of skepticism toward Lavoisier's own account of the sealed note: Guerlac's basic approach is correct and encouraging.

Another of Guerlac's fundamental mistakes is the overcoherence of his explanation, reflected in his uncritical attempt to explain Lavoisier's performance and interpretation of the litharge experiment in one stroke. As we have seen, according to Guerlac, Lavoisier performed the litharge experiment to determine whether the phenomena constituting the subject matter of the experiment could be interpreted (explained) antiphlogistically. Presumably, then, what Lavoisier learned from the litharge experiment was not much. It only "confirmed" his hypothesis. But if Lavoisier knew as much as Guerlac alleges, that is, if he knew that the metallic calxes effervesce when reduced, that metals resist calcination in closed vessels, that the calx of a metal is heavier than the regulus, and if he was also convinced that air is absorbed by a metal during calcination, what would there have been about the antiphlogistic hypothesis to test? Is it plausible that Lavoisier would waste his time with an experiment whose outcome, according to him, could only have been as predicted? And how could Guerlac account for Lavoisier's feeling of discovery which the experiment had occasioned, that is, the feeling that his discovery was one of the most interesting since Stahl? To be sure Guerlac might reply that Lavoisier's was essentially a theoretical discovery; in saying this he would be partly right. But then why does not Lavoisier deposit a sealed note on August 8, when, according to Guerlac, he had already made the theoretical discovery? What sane person would wait three months before acting on this theoretical discovery?

The truth is that Guerlac has only spuriously explained the performance of the litharge experiment. In Guerlac's account, that experiment is a useless and unenlightening one for Lavoisier. To say as much is not to account for the experiment. Most of these problems can be solved by postulating what cannot be denied, that Lavoisier learned a great deal from the litharge experiment. It will be our problem to reconstruct what he actually learned, not what he said he did.

Guerlac's frequent confusion of logical and historical facts is a third fundamental mistake. He conceives his explanatory problem to be mainly one of determining under what influences, men, facts, and ideas Lavoisier acted. He thinks there is nothing problematical about how Lavoisier came to his conclusions and experiments from the facts and ideas at his disposal because he thinks that there is nothing problematical about how *someone* (this someone is often *we now*) can be led to those conclusions and experiments from those ideas and those facts. The story of Turgot's "anticipation" is only the most explicit and self-conscious illustration of this approach. (That Turgot never carried out the experiment is no problem for this criticism; for he did "set down an identical conclusion . . . stimulated to formulate it fully and clearly by the same influences" (p. 143). But not being serious enough about these things, he never bothered to confirm it.) Guerlac's language of the "combination" and linking of factors in the mind of Lavoisier (e.g., pp. 145, 194) is only a less explicit indication. (The "combination" talk could, however, be interpreted as a sign of the incompleteness of Guerlac's account.) But to have presumably determined that Lavoisier was in possession of certain facts and ideas which might have led or can lead us to the same experiments and conclusions is at best to have solved only half of the problem. The other half is to show that Lavoisier himself could have been led from the facts and ideas known to him to his conclusion and experiments. The unquestioned fact that he did reach his conclusions and performed his experiments is no evidence that he could be led to them from ideas and facts known to him which can lead us or Turgot to them. This is true because of the differences of knowledge and information, which play a crucial role here. It is therefore self-deceptive to embark on a search for the agent's awareness of "appropriate" facts and ideas; and I suggest that this is precisely what Guerlac is doing in his search after Lavoisier's acquaintance with Hales's doctrine, effervescence in reduction and the augmentation effect. Yet I find it hard to believe that Guerlac is so deceiving himself, that he is confusing logical and historical facts. He may then be consciously trying to portray Lavoisier's thinking as that of "one of us." This would be a no less objectionable procedure.

Alternative Explanations

It is possible to restate my criticism of Guerlac's explanation in the form of a positive alternative account. This new explanation accepts most of the nonexplanatory parts of Guerlac's account, that is, the Hales-influence (L1) and Priestley-influence (L2) theses, and the nonexplanatory parts of the Mitouard episode story. But this is where my agreement with Guerlac ends. For it was in order to elaborate his mistaken theory of elements that Lavoisier originally in August 1772 planned to make certain measurements to determine the quantity of air to be found in various substances. The litharge experiment of the sealed note had been intended as one of these experiments. Indeed Lavoisier was led to experiment with phosphorus because of his connection with Mitouard and his wanting to sponsor him. At the time of these experiments he still believed in his theory of elements, though he was probably unclear about the connections between that theory and the Mitouard-motivated experiments. He was not at this time aware of the augmentation effect. He must have learned about this phenomenon toward the end of October; his mistake concerning the impossibility of the phlogistic theory being able to account for the newly established augmentation effect started him thinking, partly consciously, but mostly unconsciously, about a possible new explanation. Also sometime toward the end of October, after he had learned about the augmentation effect, he was finally led to perform some of the measurement experiments he had envisaged the previous August and which he had had to postpone in order to have the exclusive use of the Academy's great burning lens.

The results of the litharge experiments were unexpected. The opposite of what he would have predicted on the basis of the theory of elements happened. The experiment indicated that it was the calx, and not the metal, as he had believed, which contained the air. But he suddenly saw that the air given off by the calx would account for its greater weight than that of the metal. Moreover, phosphorus too had increased in weight and absorbed air. So maybe the opposite had happened in the phosphorus experiments from what had happened in the litharge experiment. To test this further, and in order not to seem to plagiarize Mitouard's work, he tried an experiment with a substance with properties similar to phosphorus, namely sulphur. The results of the sulphur experiment confirmed his hypothesis.

Those experiments were indeed important, for they invalidated his theory of elements and could be used to explain the augmentation effect, otherwise inexplicable according to Lavoisier. But in order that he would not be discredited because of his old mistaken theory of elements, he sealed the note so that he could have some time to revise his theory.

In short, the error of the theory of elements led him to perform the litharge experiment, while the mistake of regarding the augmentation effect as incompatible with the phlogiston theory led him to give the correct theoretical interpretation of that experiment.

The above alternative has its merits, but I think that in the long run, and for those who have been convinced by my criticism that Guerlac's explanation is invalid, it will be better to try a radically new approach. This approach would start with a careful thinking through of the "System of the Elements" in complete accordance with Lavoisier's ambitions expressed in its last paragraph. It would then interpret the August memorandum as a program of research intended to elaborate the theory of elements, in accordance with Lavoisier's intentions as expressed in one of the beginning paragraphs of that memorandum:

> The action of this fire, superior to that which we employ in our laboratories, has so far been applied only to metallic substances; no systematic experiments at all have been made with rocks, ores, and infinitely many mineral substances. Moreover, the few experiments which are known to us have been in open air without

introducing any variety in the operations. The experiments with the burning glass offer then a still completely new course to follow. One will become more and more convinced of this from the following reflections. [Pp. 209–10]

The third step would be to interpret the phosphorus and sulphur experiments as a study of the so-called air-absorption reactions of the theory of elements. For in the second paragraph of the draft memoir of October 20 Lavoisier says:

> Contact with open air is necessary for this operation because the vapor of phosphorus, in being converted into the volatile spirit acid of phosphorus, absorbs a small amount of air. This is proof that air is naturally contained in the composition of this mixture and that it is combined and fixed in it in the same way that happens for a large number of chemical combinations. [P. 224]

The litharge experiment would now be the real epoch-making experiment and it would be explained as somehow having forced Lavoisier to reorient his theory-of-elements program.

chapter 6

Koyré on Galileo and Descartes

I SHALL APPROACH the second volume of Koyré's *Etudes Galiléennes* in a way similar to my analysis of Guerlac's explanation, establishing first Koyré's explicit explanatory intentions, then determining what his problem is, then stating his explanatory thesis, and finally, at greater length, determining the extent to which Koyré's explanation is satisfactory.

Koyré's Claim

Koyré's explicit and principal objective in his monograph subtitled "The Law of Falling Bodies. Descartes and Galileo," is to explain something, as may be seen from his introductory remarks:

> Now, in the invention of this law, so simple that today it is
> readily understood by children, Descartes and Galileo erred
> badly. How is one to *explain* their error? . . .
> It might be objected, no doubt, that one should not look for a
> rational *explanation* of errors. . . . Errors are sufficiently *ex-
> plained* by lack of attention, distraction, or oversight. We con-
> fess to be unable to accept this objection.[1]

In saying that Koyré is trying to put forth an explanation, I am
contrasting that activity to others such as narrating, proving, refuting,
predicting, describing, or evaluating. It should not be thought, however, that explaining is incompatible with all of these activities. On
the contrary, in the course of his explanation, Koyré will be involved
in most of them. For example, he will be evaluating the scientific merits of Descartes and Galileo. The explanatory point of view, however,
is conceptually distinct. Moreover, in claiming that explanation is
Koyré's primary aim, I implicitly admit that he has others. It is obvious
that he also intends to criticize the positivist interpretation of Galileo's
scientific method.

The Problem

As evident from the above quotation, part of what Koyré wants to
explain is Galileo's and Descartes's errors in their investigation of the
problem of falling bodies. Galileo's error consisted in deriving the
correct formula from a wrong principle, whereas Descartes's consisted
in deriving an incorrect formula. More specifically, Galileo derived
the time-distance law, the proposition that the distance traveled by a
falling body varies as the square of the time elapsed, from the incorrect
principle of spatial acceleration, according to which the velocity of a
falling body would be directly proportional to the distance traveled.
In reality, the direct proportionality holds between velocity and time
or velocity and the square root of the distance. Descartes on the other
hand, when he tried to derive the time-distance relationship, derived
the following incorrect formula: a falling body travels the first half of
a given distance in three-fourths of the time required to travel that distance, that is, if the distances of fall are as 2^0, 2^1, 2^2, 2^3, etc., the times
are as 1, $4/3$, $(4/3)^2$, $(4/3)^3$, etc.; in other words, after a time of
$(4/3)^n$ units the distance is 2^n units. And this relationship makes bodies fall faster than they actually do; in fact, they fall in such a way that

if the distances are as 2^0, 2^1, 2^2, 2^3, etc., the times are as $(2^{\frac{1}{2}})^0$, $(2^{\frac{1}{2}})^1$, $(2^{\frac{1}{2}})^2$, $(2^{\frac{1}{2}})^3$, etc.

It is obvious from the third chapter and from the conclusion of Koyré's monograph that he also wants to explain why "Galileo succeeded where Descartes failed" (p. 157). For, eventually, Galileo's "attempt at deduction . . . succeeded" (p. 156); he was able "to avoid Descartes' error, and his own" (p. 157). That is, eventually Galileo derived the time-distance law from the correct principle of acceleration according to which the velocity is directly proportional to the time elapsed.

What Koyré is trying to explain then is the error of the early Galileo and of Descartes, and the success of the later Galileo in deriving a time-distance relation for falling bodies.

Statement of the Explanation

Koyré's explanation of Galileo's early error can be summarized: What led Galileo to the error of postulating the wrong principle of spatial acceleration and of fallaciously deducing from it the time-distance law was his geometricism, that is to say, his excessively geometrical attitude, his *géométrisation à outrance*. And this geometricism was the result of a natural tendency "to *visualize* in space rather than to *think* in time" (p. 96), of the intrinsically greater intelligibility of space and of spatial relations than of time (p. 97), and "perhaps above all" (p. 98) of Galileo's renunciation of causal explanation.

Koyré's explanation of Descartes's error can be briefly stated as follows. The "sources of the Cartesian error" (p. 96) are two: (1) his pure mathematician's mental attitude, and (2) his excessively geometrical outlook, or geometricism. The pure mathematician's mental attitude leads Descartes to his error indirectly by way of the following intermediaries: (1a) his inability to understand the new concept of motion brought into play by Beeckman and by the principle of conservation of motion, and (1b) his dichotomy between mathematics and physics. Descartes's geometricism, on the other hand, produces the error more or less directly.

Finally, according to Koyré, the reasons for Galileo's success are the constitutive features of his mental attitude. This attitude, which, by contrast to Descartes's purely mathematical attitude, is physico-mathematical, is characterized by the following aim, starting points, and guiding principle (p. 156). Galileo is looking for the actual properties of accelerated motion. He starts from the idea that (1) nature is mathematical, (2) the relation between theory and experience is such that theories express the essence of experience, (3) one should begin with experience, and (4) the experience with which one should begin is the descriptive mathematical laws. Finally, Galileo is guided by the idea of simplicity. In summary, Galileo ultimately succeeded where Descartes and he himself had previously failed because he paid constant and sustained attention to the real, the temporal, character of the phenomenon.

The explanation follows in Koyré's own words (p. 103):

(G1) When, in 1604, Galileo approaches afresh the problem of falling bodies, he is in possession, as we have seen [in his letter to P. Sarpi], of the formulas which relate the duration of the fall to the distance traveled; he is also in possession, as we have just pointed out, of the cardinal principle of the conservation of motion and of speed.

(G2) He renounces, on the contrary, any attempt of causal explanation and is looking for nothing but a principle, an axiom, which would allow one to deduce the descriptive laws of fall.

(G3) Now, we have equally seen, it is causal considerations which in the analysis of motion (of motion in general and of fall in particular), bring out into the foreground the concept of time. Thus it is not surprising that the renunciation of causal explanation reinforces the tendency to geometrization, and thus to spatialization.

(G4) Instead of thinking the motion, Galileo pictures it to himself. He sees the line or distance traversed with a variable velocity, and it is this line—the trajectory—which he takes as the argument of his velocity function.

(G5) The endeavor of geometrization, sustained and strengthened by the imagination, unfettered by causal thinking, goes beyond the goal which it had set for itself. The goal of dynamics was to mathematize time; now, Galileo eliminates it. The effort once accomplished led to failure, failure which Galileo does not notice at first.

(G6) For, in again going through in the inverse direction the reasoning which from the correct descriptive formulas had led him to a mistaken principle, he again finds, by starting from this principle, the right consequences from which he had begun.

The main elements of the explanation of Descartes's error are the following:

(D1) [Whenever, before his intellectual revolution of 1630, Descartes approached the problem of falling bodies] it is not a physicist, it is a pure mathematician, a pure geometer, who looks at the problem: it is for him a question of establishing the relation between two series of variable quantities. [P. 114]

(D2) Descartes is [then] a pure mathematician. That is the reason, it seems, on account of which he does not understand too well Beeckman's "principles" and gives to his question an erroneous answer. [Pp. 114–15]

(D3) [In fact,] Descartes has definitely not understood Beeckman's "principles"; moreover, he quite simply ignores the latter's intellectual conquest: the principle of conservation of *motion*. [P. 118]

(D4) [And,] not having been able—even in 1629—to understand fully the new concept of motion that the principle of its conservation brings into play, he [Descartes] is still at the stage of dissociating the causal conception and the mathematical analysis, the temporal evolution and the geometrical representation of fall. [P. 124]

(D5) [Thus,] Descartes . . . was not able to deduce the exact law of fall because he had not understood the new concept of motion that Beeckman was proposing to him and had not been able to make the physical (causal) investigation of the phenomenon coincide with its mathematical analysis. [P. 135]

(D6) [Secondly,] excessive geometrization, spatialization, elimination of time—where it cannot be eliminated—, neglect of the physical, causal aspect of the process leads Descartes—like Galileo previously and Benedetti and Michel Varron before him—to conceive uniformly accelerated motion as a motion whose speed increases proportionally to the distance covered and not proportionally to the elapsed time. [P. 115]

(D7) [In other words,] carried away by the élan of imaginative representation and of his tendency to geometrize to excess, he falls into error. [P. 119]

As for Galileo's eventual success (pp. 156–57):

(G'1) The comparison of the attempts of deduction—the one that fails and the one that succeeds—clarified by the analysis of the Cartesian texts to which we have applied ourselves, allows us to understand the reasons of the failure and those of the success.

(G'2) The thought, or, if one prefers, the mental attitude of Galileo differs noticeably from that of Descartes. It is not purely mathematical; it is *physico-mathematical*.

(G'3) Galileo does not put forth hypotheses on the possible modes of accelerated motion: what he is looking for is the real mode, the way that nature uses.

(G'4) Galileo does not start, like Descartes, from a causal mechanism in order to translate it afterwards into a purely geometrical account or even to replace it by such an account. He starts from the idea—no doubt preconceived, but which forms the groundwork of his natural philosophy—that the laws of nature are mathematical laws. *The real incarnates the mathematical.*

(G'5) Moreover, there is, in Galileo, no opposition between experience and theory; theories or formulas do not apply to phenomena from without, they do not "save" these phenomena, they express their essence. Nature responds only to questions asked in mathematical language because nature is the reign of measure and order. And if experience thus guides reasoning "by the hand" as it were, it is that, in a well conducted experiment, that is to a well put question, nature reveals its profound essence, which only the intellect is capable of understanding.

(G'6) Galileo tells us to start from experience; but this "experience" is not brute sense experience; the data to which the definition he is looking for must conform, or with which it must agree, are nothing other than the two descriptive laws—the laws of the symptoms—of fall which he already possesses.

(G'7) Galileo tells us to be guided by the idea of simplicity. But this is not merely formal simplicity: something else is in question; something analogous, no doubt, but nevertheless different: a real simplicity, as it were, a conformity internal to the essential nature of the phenomenon studied.

(G'8) This real phenomenon is motion. Galileo does not understand how it is produced, nor how—under the influence of what force—the acceleration is produced. No more than Descartes, will he, in fact, be able to profit from Gilbert's work and make use of an obscure notion—the notion of attraction—a notion which he cannot mathematize. Be that as it may, it is a real phenomenon which is in question, a phenomenon which nature produces in reality, which means: something which is produced *in time*.

(G'9) It is in that intuition, in the constant and sustained attention to the *real* character of the phenomenon that lies the reason which allows Galileo to avoid Descartes' error, and his own. Motion is, first of all a temporal phenomenon. It occurs *in time*. It is a function of time then that Galileo will try to define the essence of accelerated motion and no longer as a function of the space traversed: the space is nothing but a resultant, an accident, a symptom of an essentially temporal reality.

Criticism of Koyré's Explanation

Galileo's Error. A detailed critical exposition of Koyré's explanation may well begin by an examinination of Galileo's initial error. The error, or, to be more exact, the errors that Koyré is here trying to explain are those contained in the reasoning recorded in an undated note which was probably written prior to 1624 and possibly prior to 1609:

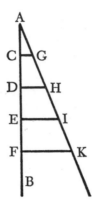

I assume (and perhaps I shall be able to demonstrate it) that the freely falling body continuously increases its velocity according to how the distance from the point of origin increases: thus, for example, if the falling body originates at point *a* and falls along line *ab*, I assume that the degree of velocity [i.e. the instantaneous velocity] at point *d* is as much greater than the degree at *c* as the distance *da* is greater than *ca*, and similarly that the degree of velocity at *e* is to the degree of velocity at *d* as *ea* is to *da*, and thus that the body has at every point of the line *ab* degrees of velocity proportional to the distances from the same points to the origin *a*. It seems to me that this principle is very natural and that it corresponds to all the experiences we have concerning those instruments and machines which operate by percussion, where the striking object has greater effect the greater the height from which it falls: and having assumed this principle, I shall prove the rest.

Let the line *ak* make any angle with the line *af*, and through the points *c, d, e, f* draw parallels *cg, dh, ei,* and *fk*: and since the lines *fk, ei, dh,* and *cg* are to each other as *fa, ea, da,* and *ca,* the velocities at the points *f, e, d,* and *c* are as the lines *fk, ei, dh,* and *cg.* Thus the degrees of velocity increase continuously at all points of the line *af* according to the increment of the parallels drawn from all those same points. Moreover, since the velocity with which the moving body has come from *a* to *d* is made up of all the degrees of velocity acquired at all points of the line *ad,* and since the velocity with which it has passed the line *ac* is made up of all the degrees of velocity acquired at all points of the line *ac,* it follows that the velocity with which it has passed the line *ad* is to the velocity with which it has passed the line *ac* in the same ratio that all the parallel lines drawn from all points of the line *ad* to line *ah* are to all the parallels drawn from all points of the line *ac* to line *ag;* and this ratio is the same as that between the triangle *adh* and the triangle *acg,* that is between the square of *ad* and the square of *ac.* Thus the velocity with which the line *ad* has been traversed is to the velocity with which the line *ac* has been traversed in a ratio double that between *da* and *ca.* And *since the ratio between velocities is the inverse of the ratio between times* (in that to increase the velocity is the same as to diminish the time), it follows that the time of the motion *ad* is to the time of the motion *ac* in a subduplicate ratio of that of the distance *ad* to the distance *ac.* That is, the distances from the origin of motion are as the squares of the times, and, dividing, the distances covered in equal times are as the odd numbers from unity: which corresponds to what I have always said and observed in experiments; thus all the truths agree with each other.

And if these things are true, I show that the velocity in violent [upward] motion decreases in the same way in which, in the same straight line, it increases in natural motion [free fall].[2]

Koyré finds a double error in Galileo's reasoning:

It is true, no doubt, that the ratios of the velocities are inverse those of the times; on condition that the basis of comparison, that is, the distance covered, be the same and not, as in our case, different. And it is also true that the total velocity of the moving body is the sum of the (instantaneous) velocities acquired at every point of the way; as it is equally the sum of the velocities acquired at every instant of the motion. But these "sums" are not similar; the constant and uniform increase with respect to time will not be constant and uniform with respect to space and vice-versa; in particular the "sums" of velocities which increase as a linear function of distance traversed cannot be represented by triangles. This representation is valid only for an increase uniform with respect *to time.* Once more Galileo geom-

etrizes excessively and *transfers to space what holds for time*.
[P. 106]

But Koyré's claim that Galileo was led to these errors by excessive geometrization is unacceptable. One error in Galileo's reasoning is his use of the proposition that "the ratio between velocities is the inverse of the ratio between times." This is to say that velocity is inversely proportional to time. And, as Koyré points out, this is true for a fixed distance and false if, as in free fall, different distances are involved. What Koyré does not point out, however, is that this obviously amounts to ignoring spatial considerations. Far from having geometrized to excess, in this instance, Galileo has not geometrized at all. He has neglected the geometry of the situation altogether. Insofar as geometrization is involved at all, then, it was *lack* of geometrization that led Galileo to the present error.

Galileo's other mathematical error, according to Koyré, is his use of a triangle representation which is valid only for an increase uniform with respect to time and not for one uniform with respect to space; that is, Galileo "transfers to space what holds for time." But if this is so, Galileo is temporalizing space, i.e., attributing to space a characteristic of time. And excessive geometrization cannot lead to this sort of error; it can only lead to representing quantities geometrically that ought not to be so represented, or inappropriately projecting spatial, and not temporal, properties onto quantities. The thesis (G5) is thus unacceptable.

(G5) remains problematic even though a third error committed by Galileo, his postulating the wrong principle of spatial acceleration, could have stemmed from geometricism. For, on the one hand, Koyré is right in saying:

> Instead of thinking the motion, Galileo pictures it to himself. He sees the line or distance traversed with a variable velocity, and it is this line—the trajectory—which he takes as the argument of his velocity function. [P. 103]

But, there is a problem here with the ultimate source of the error; with the reason for Galileo's geometricism itself. Koyré's reason is that Galileo was bent on mathematizing nature and that prior to Descartes's creation of analytical geometry any mathematization had to be geometrization. In Koyré's own words, one would like to know why, of two equivalent relations, or at least of two relations which they believe to be such (velocity proportional to time elapsed, and velocity proportional to distance traveled), why do Leonardo, as well as later Galileo and Descartes, resolutely choose the second. The reason for it seems to us very profound and very simple at the same time. It lies entirely in the role played in modern science by geometrical considerations, by the relative intelligibility of spatial relations.

The process from which modern physics has emerged consists of an effort to rationalize, that is to say to geometrize space and to mathematize the laws of nature. In reality it is the same effort which is in question, since to geometrize space is nothing but to apply geometrical laws to motion. And *how—before Descartes—could one mathematize something if not by geometrizing it?* [P. 96, my italics]

Koyré's connection between the unavailability of analytic geometry and the present error of Galileo is plausible; it is, however, unacceptable because, as Koyré himself tells us, the creator of analytic geometry himself will be a victim of geometricism, whereas Galileo will eventually overcome it without analytic geometry. Thus Galileo's physical error is no more accounted for by the above "profound" reason than his two mathematical errors are by his geometricist attitude.

Another source of Galileo's error, in Koyré's opinion, was his renunciation of causal explanation. The problem here is that, even if this renunciation could account for at least some aspects of Galileo's error, Koyré gives no shred of evidence that Galileo had, at the time of his error, made the renunciation. In fact, when Koyré, in his analysis of Galileo's letter to P. Sarpi (p. 88), first attributes this renunciation to Galileo, he refers to his later discussion (p. 146), in the context of his explanation of Galileo's success, quoting a passage from Galileo's *Discourse*. The passage does indeed indicate Galileo's abandonment of the causal explanation of accelerated motion, but at the time Galileo wrote the *Discourse*, sometime between 1633 and 1638, not the time of Galileo's earlier unsuccessful investigations, sometime between 1604 and 1624.

But Koyré's claim that Galileo had renounced the causal explanation of accelerated motion by 1624 at the latest, is worse than groundless or inappropriately grounded. It is probably false. For if there is a connection between the renunciation and the outcome of Galileo's investigation of the problem of falling bodies, the likely connection is with Galileo's success, for it is at the time of this success that the renunciation is demonstrably present. It is likely, therefore, that at the time of Galileo's error he had not yet abandoned causal explanation. Our general conclusion is that, though Koyré's account is not without value, the sources of Galileo's present errors cannot be those alleged by Koyré's explanation.

Descartes's Error. Let us now examine Koyré's explanation of Descartes's error. Descartes made the error of deriving the wrong time-distance relationship on at least four different occasions, all before 1630. The first time was in his "private thinking" when the question of the time-distance relationship was initially brought to his attention by Beeckman; the mistake can now be found in a note or fragment which is part of what the editors of Descartes's works have collected under the heading of "Private Thoughts." The second time was in Descartes's formal reply to Beeckman's question; this derivation is contained in a document which is not exactly a letter from Descartes to Beeckman but a paper prepared by Descartes for Beeckman with a certain René du Perron as an intermediary. The third and fourth time that Descartes repeated the error was in his correspondence with Mersenne. This can be found in two different letters from Descartes to Mersenne, the second letter being an attempt to clarify some points in the first.

The derivations in these four documents vary as to their completeness and their method. They all agree, however, on one point: all derive the wrong relation between distance and time, that is, the relation that bodies fall a distance of 2^n units after a time of $(4/3)^n$ units, which is faster than they actually do.

In order to see whether Koyré's explanation can account for Descartes's error, we have to follow in detail the reasoning by which Koyré tries to connect Descartes's error to his pure mathematical attitude and his geometricism. That Koyré is right in claiming that Descartes did approach the problem of falling bodies with a pure mathematician's attitude may be granted upon an examination of Koyré's evidence, the above-mentioned note, in which is recorded the first of Descartes's "private thoughts" on the matter:

> Few days ago I happened to become acquainted with a very intelligent man [Isaac Beeckman] who asked me the following question:
> He says: a stone falls from A to B in one hour; it is perpetu-

ally pulled by the earth with the same force and loses none of the velocity which was impressed to it by the preceding attraction. Now, *according to him,* whatever moves in the void moves eternally. It is asked how long it takes to cover a given distance.

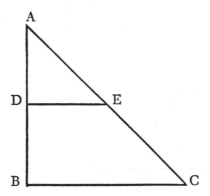

I have solved the problem. In the right isosceles triangle, ABC represents the space (the motion); the inequality of the space from point A to the base BC, the inequality of the motion [i.e. the variation in the velocity]. Consequently, AD will be passed over in the time represented by ADE, and DB in the time represented by DEBC: where it is necessary to note that the smaller space represents the slower motion. But ADE is a third of DEBC: consequently, AD will be passed over three times more slowly than DB.

But one can also put this question differently: assume that the attractive force of the earth is equal to that of the first moment and that a new one is produced as long as the previous one persists. In that case the problem will be solved by the pyramid [i.e. the second half of the distance, namely DB, will be covered in (⅔)³ of the time]. [Pp. 113–14] [3]

We may then agree with Koyré that on this matter "Descartes is a pure geometer, a pure mathematician" (p. 113). The thesis (D1) thus seems to be true. We may also agree with Koyré in saying that Descartes did not at the time understand the new concept of motion (p. 118, and (D3) above). Descartes's formal reply to Beeckman's question supports this conclusion:

To the question asked, where one supposes that a new force is added [to that] with which the body tends toward the bottom, I say that this force increases in the same way in which increase the transversal lines *de, fg, hi* and the infinitely many transversal lines which can be imagined between them. And if I were to prove it, I would regard the square *alde* as the first minimum or degree of motion caused by the first attractive force of the earth that one can imagine. For the second minimum of motion we would have the double, namely *dmgf*: indeed the first force which was present in the first minimum remains and another one, new and equal to the previous one, is added to it. Similarly in the third minimum of motion there would be three forces, namely those of the first, second, and third minimum of time, etc. Now this number is triangular, as I will elsewhere explain perhaps at greater length, and it will appear to represent the

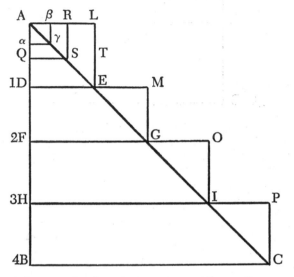

figure of the triangle *abc*. However, you will say, there are some extra portions, *ale, emg, goi*, etc., which protrude from the figure of the triangle. Consequently the figure of the triangle cannot express the progression in question. Now I answer that these extra portions are due to the fact that we have given an extension to these minima which it is necessary to imagine as indivisible and not as being made up of any parts, which can be proved in the following manner. I divide the minimum *ad* into two equal parts at *q*; then *arsq* will be the (first) minimum of the motion, and *qted* the second minimum of the motion, in which there will be two minima of forces. Similarly we will divide *df*, *fh*, etc. Then we will have the extra portions *ars, ste*, etc. They are smaller than the extra portion *ale*, as it is obvious. Let us go further. If I take a smaller minimum, such as *aα*, the extra portions will be even smaller, such as *aβγ*, etc. Finally if for minimum I take the real minimum, the point, then these extra portions will be nothing, since they will not be able to constitute the whole point but evidently only half of the minimum *alde,* and half of a point is nought. [Pp. 116–17] [4]

Now, Descartes's incomprehension will be apparent if we admit with Koyré that the new concept of motion, and the one that Beeckman is using, "consists: (a) in a *clear* affirmation of the law of the conservation *of motion* which is thus freed of the conception of *impetus;* (b) in the elimination of all cause internal to the movable body" (p. 116, n. 1). Whereas the above Cartesian argument replaces the concept with that of internal force, as Koyré says:

> He starts with the idea that the *velocity* is proportional to the *force* and concludes from it that a constant force produces a constant velocity. Thus he falls back into the classical conception of the *impetus* physics. He thinks that if the falling body accelerates its motion, it is because it is more strongly pulled by the earth at the end of its movement than at its beginning, or, to speak his language, because the attractive force of the earth produces in the stone a growing motor force: thus he adds . . . the acting forces and not simply the velocities. [P. 118]

But though we may grant Koyré the reality of Descartes's pure mathematician's attitude and of his incomprehension of the new concept of motion, we cannot accept them as sources of Descartes's error. For his pure mathematician's attitude no more produced his incomprehension than did the latter produce his error (pp. 114–15, 135, and (D2) and (D5) above).

The connection between Descartes's pure mathematicism and his lack of comprehension is not immediately apparent and cannot be accepted without argument; for what is in question is the possibility of the connection and not just its reality. Koyré does have an argument:

> . . . the concept of motion which Beeckman—implicitly—puts forth (it is the concept of motion of classical physics) lies somehow in the narrow frontier between the mathematical (the geometrical) and the physical (the temporal). . . . [Thus] motion, this paradoxical entity which is a state of a movable thing and which, nevertheless, passes from one movable thing to another, which incarnates change and which, at the same time, remains identical to itself, this motion seems to him [Descartes] a bastard entity. And voluntarily as well as instinctively he replaces [in his reasoning in the Formal Reply] this concept by

> those, more solid—and clearer, more easily *imaginable*—of mo-
> tor force on the one hand and trajectory on the other. [P. 119]

Despite the argument it is not clear what Descartes's pure mathematicism has to do with his incomprehension of Beeckman's "motion."
In fact, any plausibility that Koyré's argument may have is due entirely to his apparently parenthetical mention of the superior imagin-
ability of Descartes's concepts to Beeckman's. But then what leads
Descartes to his incomprehension is the fact, mentioned by Koyré
himself in a footnote to the word *imaginable* in the above quoted
passage, that "Descartes' physics is, alas, an *imaginative* physics, and
often, in physics, a clear concept is for him only one clearly imagined"
(p. 119, n. 1). But then the connection is between Descartes's geometricism and his incomprehension, which is stated in (D6) and (D7)
and which we shall examine later. Thus any connection between Descartes's pure mathematicism and this incomprehension is mysterious
and cannot be used to explain anything.

However, as we have mentioned above, Descartes did fail to understand Beeckman. Whatever accounts for this incomprehension, the
possibility remains that it produced Descartes's error, which is what
Koyré claims in (D2) and (D5). The possibility will not remain open
for long, however; in particular it will not survive the following considerations, made by Koyré himself, who does not seem to realize
their consequences, namely, the refutation of (D2) and (D5). In fact,
the Cartesian error is an

> error which, curiously, even with his physics of force, he could
> have, in principle, avoided. (All he had to do was to maintain
> strictly the parallelism between force and velocity and to con-
> tinue to think causally, that is as a function of time.) If he falls
> into the error, it is because, by substituting the trajectory for
> the motion, he makes the trajectory—and not the time—the ar-
> gument of his function. [P. 119]

Koyré has fallen back, once again, into geometricism as the source
of Descartes's error. He has done this by pointing out that the new
concept of motion is irrelevant to that error. Since Descartes talks
about forces instead of velocity, he could have integrated the forces
over time. If he had correctly done this, he would have found the
"sum of the forces" to vary as the square of the time. And since for him
velocity is proportional to force, he could have taken the "sum of the

forces" to be a measure of the "motion" accomplished; so that the motion (distance) would have been seen to vary as the square of the time. Thus Descartes's incomprehension of Beeckman's concept could not, on the evidence, have produced the error.

It is perhaps easy to see how a pure mathematician's attitude would result in a mathematics-physics dichotomy. What is unclear, in fact untenable and unlikely, is that Descartes's mathematics-physics dichotomy can account for his error, which Koyré claims in (D5). We may begin by accepting Koyré's claim that Descartes did divorce mathematics and physics (D4). This claim Koyré supports by his analysis of the formal reply to Beeckman quoted above. What Koyré wants to note about that reply is that

> it is difficult to imagine a text which, like this one, would combine a supreme mathematical elegance with the most irretrievable physical confusion. . . . Yet he succeeds brilliantly in his mathematical deduction. . . . It is when he tries to translate the results of his integration in terms of space . . . that he falls into the error. [Pp. 118–19]

Koyré seems to be saying that there are two main elements in Descartes's formal reply, the successful solution of a pure mathematical problem of integration, and the unsuccessful application of the mathematical results to a physical situation. The mathematical problem that Descartes seems to have successfully solved is that the sum of the area of rectangles of the type ALED, DMGF, FOIH, and HPCB approaches the area of the triangle ACB as the rectangles become narrower and narrower.

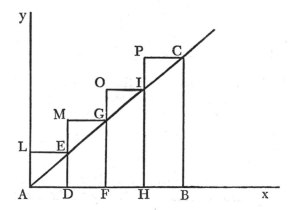

He is unable to apply this result correctly to the problem of falling bodies. If we accept Koyré's interpretation it is true that when Descartes attempts to cross the gap from mathematics into physics he fails. But does he fail because he dissociates mathematical from physical analysis, as Koyré also claims (p. 115, and (D5) above)? Not on Koyré's own evidence, for when Descartes applies mathematics to physics what happens is that he is "carried away by the élan of imaginative representation and of his tendency to geometrize to excess" (p. 119). Once more, instead of actually connecting Descartes's error to some other identifiable aspect of his mental attitude, Koyré falls back on geometricism. The only source of the Cartesian error left is then geometricism.

Can Koyré explain Descartes's error by geometricism as he says in (D6) and (D7)? The only argument he gives in support of these claims is his analysis of the third paragraph of the note recording Descartes's private thoughts quoted above (pp. 96–97).

> The line ADB—which for Beeckman represents the time elapsed —represents for him [Descartes], naturally, the traversed trajectory. And the problem is transformed: a trajectory is traversed with a uniformly variable velocity; the problem is then to determine the velocity at each point of the way. The triangles ADE, and ABC, which for Beeckman represent the space passed over (the distance), represent for Descartes the motion of the movable body, that is the "sum of the velocities" realized. And, very plausibly, he concludes: the "sum of the velocities" being triple, the distance DB will be passed over three times faster. Time is recovered, but too late; excessive geometrization, spatialization, elimination of time—where one cannot eliminate it—, neglect of the physical, causal aspect of the process lead Descartes—like Galileo previously and Benedetti and Michel Varron before him —to conceive uniformly accelerated motion as a motion in which the speed increases proportionally to the distance covered and not proportionally to the time elapsed. [P. 115]

Now, both Koyré's translation of the original Latin text, and his interpretation of his own translation are puzzling enough; his conclusion is completely wrong. It accounts, in fact, for an error that Descartes did not make, "to conceive uniformly accelerated motion as a motion in which the speed increases proportionally to the distance covered and not proportionally to the time elapsed." Koyré gives no evidence that Descartes made this error. Nor can it be found either in the previously mentioned note or in any of the three other documents in which Descartes derives the wrong time-distance relation. Descartes's thinking is in fact far too confused for him to be able to make Galileo's physical error. Insofar as there is any indication of Descartes's thinking on the matter, Descartes "imagines that at *each instant* [which instants he seems to have numbered in his diagram (p. 117, and p. 98 above)] is added a new force with which the falling body tends toward the bottom . . . [and] that this force increases in the same way in which increase the transversal lines *de, fg, hi,* and the infinitely many other transversals which can be imagined between them" (p. 116, my italics). If Descartes is conceiving the velocity to be proportional to anything, he is conceiving it to be proportional to the force and in turn to the time, since the force is for him proportional to the time. The evidence of Descartes's second letter to Mersenne is even more compelling:

> the velocity is impressed by gravity as *one* in the first moment, and again as *one* in the second moment by the same gravity, etc. Now, *one* at the first moment, and one at the second make *two,* and *one* at the third make three, and so [the velocity] increases in an arithmetical proportion. [P. 126]

The velocity increases in an arithmetical proportion: in arithmetical proportion relative to what? Obviously relative to the "moments" or time that Descartes has just been talking about. Thus, not having made the error, Descartes could not have been led to it by geometricism. Just as in the explanation of Galileo's "error," Koyré has again been careless about what the error is that he is trying to explain. In particular, Descartes and Galileo made no common error as Koyré claims in the introduction; there was no "coincidence . . . in error" (p. 84).

We have just shown that even if geometricism had characterized Descartes's attitude it certainly would not have led him to an error he did not make. It is also questionable to say that Descartes had a geometricist attitude in this context. Koyré supports his attribution of this attitude to Descartes by his interpretation of Descartes's note. Unfortunately, Koyré's interpretation of the note completely distorts it. First of all, Koyré misinterprets his own translation of it at the crucial point where temporal considerations are in question. Descartes, in his statement of the solution to the problem, says that "AD will be passed over in the time represented by ADE" thus clearly implying that the triangle ADE represents time; whereas Koyré "interprets" this by saying that "the triangles ADE and ABC . . . represent for

103

Descartes the motion of the moving body, that is the 'sum of the velocities' realized" (p. 115, and p. 102 above). No wonder that Descartes "eliminates time." Koyré has eliminated it for him.

Unquestionably, Descartes is confused and even incoherent in his thinking about time. But it is one thing to think unclearly and incoherently about time, and it is another to eliminate it, in accordance with a geometricist attitude. This statement should be sufficient proof of Descartes's confusion and incoherence: "Consequently, AD will be passed over in the time represented by ADE, and DB in the time represented by DEBC: where it is necessary to note that the smaller space represents the slower motion" (p. 114).

Koyré's translation has Descartes saying in the same sentence that ADE and DEBC represent two different things, namely, the times and the motions corresponding to the distances AD and DB respectively. Moreover, if Descartes is saying that ADE and DEBC represent the respective times, since he also says that ADE is a third of DEBC, he could not have failed to conclude that AD, the first half of the distance, is covered three times as fast as DB, the second half. And this conclusion flatly contradicts Descartes's main explicit conclusion.

One begins to suspect that the confusion is in Koyré's mind rather than Descartes's; and one is led to look at the original text to see whether the translation is faithful. All the more so since Koyré's translation of the passage involves a semantical ambiguity in the term *space,* the French *espace.* In his translation Koyré uses this term to refer both to "motion," and to "space" in the sense of "geometrical space enclosed by the figures drawn in the diagram," (p. 115, and p. 97 above).

The semantic inconsistency is easier to correct than the logical one. The wording of the original Latin text accompanying the diagram is as follows:

> Solvi quaestionem. In triangulo isoscelo rectangulo, ABC spatium ⟨motum⟩ [*sic*] repraesentat; inaequalitas spatii a puncto A ad basim BC, motus inaequalitatem. Igitur AD percurritur tempore, quod ADE repraesentat; BD vero tempore quod DEBC repraesentat: ubi est notandum minus spatium tardiorem motum repraesentare. Est autem AED tertia pars DEBC: ergo triplo tardius percurret AD quam DB.[5]

In translating the second sentence of this passage as "In the right isosceles triangle, ABC represents the space (the motion)," Koyré is taking the word *spatium* to be the object of the verb *repraesentat* and the word *motum* as synonymous with *spatium*. However, if the word *spatium* is to be regarded as used in the same sense there as in the rest of the passage, then it must be the subject of the sentence, while *motum* becomes the only object of the verb. Thus we get "in the right isosceles triangle, the space ABC represents the motion."

The elimination of the logical inconsistency is not so easy; it does not involve such an obvious correction of Koyré's translation. The following translation of mine does, however, remove the logical inconsistency:

> I have solved the problem. In the right isosceles triangle, the space ABC represents the motion: the variation in the space from point A to the base BC the variation in the motion. Therefore, AD is passed over in a certain time, and this [motion from A to D] is represented by ADE; so also is DB, and this [motion from D to B] is represented by DEBC: where it must be noted that the smaller space represents the slower motion. But ADE is a third of DEBC: thus AD is covered three times as slowly as DB.

Thus Koyré's claim that geometricism led Descartes to commit Galileo's physical error (D6) is doubly false, first because Descartes did not commit that error and then because geometricism or neglect of time does not characterize Descartes's attitude in this context. The error that Descartes does commit in the above note is simply that of deriving and not assuming a wrong velocity-distance relation. This relation is expressed by saying that the average velocity increases by a factor of 3 in the second and final half of the distance fallen by the body; and this relation is very different from the principle, also wrong, that the instantaneous velocity is directly proportional to the distance fallen. And the immediate source of the error is that Descartes represents the motion (speed) of the body by the area of a surface, the figures in the diagrams. The collapse of (D6) and (D7) means the collapse of Koyré's explanation of Descartes's error.

Finally, this explanation is incomplete in that it generates the problem of why Beeckman misinterpreted Descartes's reply in such a way as to give a correct derivation of the time-distance law, without detecting Descartes's error. Koyré does account for Beeckman's success by saying that "Beeckman, good physicist no doubt, is a very mediocre mathematician" (p. 122). This explanation is unsatisfactory, especially in the context of Koyré's approach, outlined in his introduction, of attempting to give a rational explanation of errors.

Galileo's Success. Galileo succeeded in formulating the correct principle of temporal acceleration and in deriving from it the time-distance law. What led Galileo to this success was, according to Koyré, his physico-mathematical attitude. The evaluation of this part of Koyré's explanation is somewhat complicated in that support for it does not consist of specific arguments or evidence for each of the claims (G′1)–(G′9) constituting the statement of the explanation. Rather, the support consists of an account of Galileo's formulation of the principle of acceleration and derivation from it of the time-distance law, an account emphasizing the aspects relevant to the relation between theory and experience. Then by way of general conclusion Koyré states, without direct or explicit argumentative justification, what he judges to be the reasons for Galileo's success (pp. 156–57).

In (G′3), Koyré states the first element of Galileo's mental attitude responsible for his having succeeded where Descartes and he himself had previously failed: "Galileo does not put forth hypotheses on the possible modes of accelerated motion: what he is looking for is the real mode, the way that nature uses" (p. 156).

First of all, however, Galileo *does* put forth hypotheses on the possible modes of accelerated motion. Galileo's own definition of uniformly accelerated motion (as motion whose instantaneous velocity increases directly with time) is an hypothesis about a possible mode of accelerated motion which Galileo is convinced corresponds to the essence of the naturally accelerated motion of falling bodies.

Second, Koyré clearly needs a contrast in attitudes since he is trying to contrast the reasons for Galileo's success and those for Descartes's failure. But the contrast described by him in (G′3) is spurious because there is no conflict between putting forth hypotheses about possibilities and looking for what is real. Far from conflicting, these two elements are probably related as means and end, respectively.

In particular, and this is a third objection, both elements characterize Descartes's attitude as well. Koyré is trying to characterize a contrast between Descartes's pure mathematical and Galileo's physico-mathematical attitudes. And the obvious implication is that Galileo does not, *like Descartes,* put forth hypotheses, and what he (Galileo) is looking for, *unlike Descartes,* is the real. But Descartes, no less than Galileo, is looking for the real properties of the motion of falling bodies. In fact, he is constantly aware of the need to correct Galileo's results, which at best hold only for ideal conditions (perfect vacuum, mass points, etc.) if one is to consider the real case. For example, Descartes writes to Mersenne:

> Regarding what you send me of Galileo's calculation of the velocity with which falling bodies move, it does not in the least agree with my philosophy, according to which two lead globes, for example, one of one pound and the other of one hundred pounds, will not have the same ratio [of velocities] between them as two wooden ones, one also of one pound and the other of one hundred pounds, nor the same as two other lead ones, one of two pounds and the other of two hundred pounds. All these are things which he does not distinguish at all, which fact makes me think that he cannot have attained the truth. [Pp. 133–34]

And it is not only Galileo's results that Descartes judges to be in need of appropriate corrections, but his own as well. On a previous occasion, he wrote to Mersenne that "if you let fall a ball *in vacuo* from a height of 50 feet, regardless of the substance of which it is made, it will take always exactly three times as long for the first 25 feet as for the last 25. *But in the air it is an altogether different affair . . .*" (pp. 126–27, my italics). Indeed, Descartes's aim is "to determine the velocity of a falling stone, not at all *in vacuo*, but *in the real atmosphere*" (p. 134).

Finally, Koyré's claim is misleading in that it suggests a positivist interpretation of Galileo's general attitude, namely, that Galileo "frames no hypotheses" but is just after the facts. Such an interpretation is both false and incompatible with Koyré's.

But if Koyré's claim is misleading in the above sense, should not one find a charitable interpretation of it which is not misleading and preserves the intended contrast? One such interpretation would be: Galileo is not, like Descartes, dealing with speculative possibilities ("feigning" hypotheses) but is interested, unlike Descartes, in the actual physical problem. Unfortunately, even this interpretation would not be a correct characterization of Descartes's attitude. For though Descartes was dealing with speculative-mathematical possibilities at the time of his private thoughts about, and formal reply to, Beeckman's question in 1618, he is clearly not doing that in his 1629 letters to Mersenne (pp. 123–27). And these letters are prior to what Koyré regards as Descartes's intellectual revolution in 1630, after which time the characterization would be partly correct. That is, it would presumably be accurate to say that after 1630 Descartes was dealing with speculative possibilities and was uninterested in the actual physical problem to the extent that he had lost interest in the physical problem. He had done so, in fact, because of the astronomical complexity of the problem from his new point of view, namely, "to proceed 'according to the order of reasons' and not according to that of the subject matter" (p. 127). Thus whether literally or charitably interpreted, Koyré's claim (G′3) must be rejected.

Koyré's claim concerning the second element of Galileo's physicomathematical attitude (G′4) may be accepted as a characterization of that attitude, but the contrast with Descartes's does not hold. In particular it is not clear why Descartes's causal mechanism is regarded as a purely geometrical or intemporal framework if, as Koyré claims in his explanation of Galileo's error, "it is causal considerations which in the analysis of motion (of motion in general and of fall in particular) bring into the foreground the concept of time" (p. 103), so much so that Koyré seems to be saying that to think causally is to think temporally.

As for the third element of Galileo's attitude mentioned in (G′5), Koyré's characterization of it may be accepted. But this attitude of Galileo's contrasts not to anything that Koyré has attributed to Descartes but to the traditional Aristotelian attitude. In fact, in Koyré's narrative account, if one looks for what might serve as evidence for (G′5) one finds the Aristotelian and the Galilean points of view contrasted. After Galileo has derived, in the *Discourse*, the special properties of uniformly accelerated motion from the basic definition of uniform acceleration, the Aristotelian Simplicio asks whether such is the acceleration that nature actually uses in the case of falling bodies.

> "Very reasonable demand," Galileo thinks, "and in accordance with custom in the sciences which apply mathematical demonstrations to conclusions concerning nature; (that is the case, for example, with perspective, astronomy, mechanics, music, etc.); in these subjects the authors ask for a correspondence with experience as the confirmation of their principles, which are the foundation of all their subsequent construction."
>
> The agreement between the Galilean and the Aristotelian seems complete. But, in fact, the same words have a profoundly different meaning. What Aristotelian empiricism requires is "experience" which may serve as a basis and foundation for the

theory; what Galileo's epistemology, *apriorist* and experimentalist at the same time (one could even say: the one because of the other), offers to him is experiments constructed from the point of view of a theory and whose role is to confirm or disconfirm the application to reality of laws deduced from principles the foundation of which is elsewhere. [P. 153]

But if the present aspect of Galileo's attitude offers no contrast to Descartes's, it is not clear how it can contribute to the explanation of Galileo's success as contrasted to Descartes's failure.

In his next two claims, (G′6) and (G′7), in which he states that Galileo tells us that one should begin with experience and be guided by the idea of simplicity, Koyré abandons completely the attempt to contrast Galileo's attitude to Descartes's. With it we may take Koyré to have abandoned the attempt to explain Galileo's eventual success in contrast to Descartes's error. Let us then take Koyré to be trying to explain Galileo's success per se and ask whether Koyré's characterization of Galileo's mental attitude, which may be accepted, can account for that success.

Our answer must be a categorical negative. Galileo's practical success of having achieved certain scientific results cannot have been produced by his mental attitude but only by his practical attitude. It must be due primarily to what Galileo actually does and not to what he says or believes. In fact, the elements of the mental attitude attributed by Koyré to Galileo are elements of Galileo's theory of scientific method, and not elements of his actual scientific procedure. They are ideas about the mathematical character of nature, the relation between theory and experience, and the importance of simplicity. I think it makes little sense to claim that someone succeeded in doing *x* because he had certain ideas or said certain things about how and why things like *x* should be attained. At best this someone succeeded in doing *x* because he had certain ideas about how and why things like *x* should be attained, and behaved accordingly. Practical success, even intellectual practical success, cannot be achieved without practical effort, or practical intellectual effort, as the case may be. It is Galileo's intellectual practice that must be examined if one wants to find the sources of his success.

That Koyré is not entirely unaware of this is shown in his narrative account (in his own practice), where, before he states his conclusion, it is Galileo's methodological practice that he examines primarily.

Moreover, he states a conclusion about it (G′9) though he thinks it to be just a concise summary of his conclusions about Galileo's "mental attitude."

(G′9) attributes to Galileo a constant and sustained attention to the real character of the phenomenon of fall. Koyré's evidence for (G′9) is the alleged care Galileo shows in attributing to real falling bodies the properties which he has derived in his geometrical demonstrations. For example:

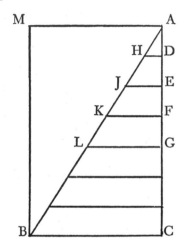

The demonstration in the *Dialogue* is based on the continuity of acceleration and makes use of the notion of "instantaneous velocity" (degree of velocity) and of "sum of velocities," which is identified with the space traversed. "In accelerated motion," Galileo tells us, "the velocity increase is continuous and . . . the degrees of velocity, changing from moment to moment . . . are infinite; also we will be able to illustrate our conception better by drawing a triangle ABC, marking on its side AC as many equal parts as one wishes, AD, DE, EF, FG, and drawing through points D, E, F, G straight lines parallel to the base BC; next I ask that one imagine that the parts of the line AC are equal times; that the parallels drawn through points D, E, F, G represent the degrees of velocity, which are accelerated and growing equally in equal times; that point A is the state of rest from which the body starts falling, which in time AD will have acquired the degree of velocity DH; that in the next period the velocity will have increased from the degree DH to the degree EJ and afterwards in the successive periods will become greater according to the increase of the lines FK, GL, etc. Now, since the acceleration takes place in a continuous manner from moment to moment, and not discretely from one period to the other, and since the origin A represents the moment of minimum velocity, that is to say, the state of rest and the first instant of the subsequent time AD, it is clear that before the attainment of the degree of velocity DH, which occurs in time AD, the falling body will have passed through an infinity of other still smaller degrees, attained during the infinite number of instants which are in the period DA and which correspond to the infinite number of points which are on the line DA; moreover, in order to represent the infinity of degrees of velocity which precede the degree DH, one must imagine an infinity of smaller and smaller lines drawn from the infinity of points of the line DA and parallel to DH; and this infinity of lines will finally represent the surface of the triangle ADH. It is thus that we will represent the whole distance traversed by the falling body with a motion which, beginning from rest and being accelerated uniformly, will have attained an infinity of degrees of increasing velocity, corresponding to the infinity of lines which, beginning at point A, are supposed to be drawn parallel to HD and to JE, KF, LG, and BC; and the motion may be continued for as long as one wishes. Now, let us complete the parallelogram AMBC and extend to its side BM not only the parallels drawn in the triangle but also the infinitely many that can be imagined as being drawn from all points of AC. Then BC will be the longest of the infinitely many lines of the triangle and will represent the greatest degree of velocity acquired by the falling body in its accelerated motion, and the surface of the triangle will be the sum-total of all the velocities with which it will traverse a certain space in the time AC; similarly the parallelogram will represent a sum-total of as many degrees of velocity, but equal each to the maximum velocity BC; and this sum-total of velocities will be double the sum-total of the increasing velocities of the triangle, just as the parallelogram is twice (the area of) the triangle; thus, if the falling body, which has employed the degrees of accelerated velocity corresponding to the triangle ABC, has passed over a certain distance in a certain time, it is rather verisimilar and probable that, in employing the uniform velocity corresponding to the parallelogram, it will cover in the same time and with a uniform motion a distance double that which was covered by it in its accelerated motion." [Pp. 147–48] [6]

Koyré takes Galileo's qualification "it is rather verisimilar and probable," which occurs in the last sentence, as evidence of Galileo's "sustained and constant attention to the *real* character of the phenomenon" (p. 157) of fall. But Koyré's interpretation is seriously inaccurate. For what Galileo is qualifying is not his conclusion per se but his process of reasoning. For the proposition which follows the final adverb *thus* is actually an unqualified hypothetical of the form "If A, then it is probable that B." What is being qualified is the consequent of the conclusion. What Galileo is uncertain about is whether, even if the falling body has employed the degrees of accelerated velocity corresponding to the triangle and has passed over a certain distance in a certain time, it really follows that in employing the uniform velocity corresponding to the parallelogram the body will cover in the same time double that distance. Galileo's procedure does not support Koyré's claim but rather the proposition, irrelevant for his purposes, that Galileo paid constant and sustained attention to the validity of his reasonings.

Our interpretation would remain unaffected were we to say that Galileo's qualification does apply to his conclusion, because his conclusion is really the proposition that "it is rather verisimilar and probable that, in employing the uniform velocity corresponding to the parallelogram, it [the movable body] will cover in the same time and with a uniform motion a distance double that which was covered by it in its accelerated motion" (p. 148). And this proposition is the "real" conclusion because the antecedent of the hypothetical is being tacitly affirmed, so that the "thus, if A, then it is probable that B" really means "thus, since A, probably B." Our interpretation would remain unaffected because Galileo's reason (his present, though perhaps not his only, reason) for the "probably" is that he thinks that B follows only probably from his assumptions and not that he thinks it applies to reality only with probability.

This interpretation, validly reached from an examination of Koyré's own argument and evidence, is supported by Galileo's original text, which states unmistakably that the qualification applied to the reasoning. In fact, in a passage which immediately follows, Galileo has Sagredo, the intelligent layman and neutral thinker in the *Dialogue,* say that Galileo is over-cautious in calling his own reasoning probable. Sagredo finds the argument necessarily valid: "I am entirely persuaded. But if you call this a probable argument, what sort of thing would rigorous proofs be? I wish to Heaven that in the whole of ordinary philosophy there could be found even one proof this conclusive." [7]

The original text, besides confirming our interpretation, provides us with further evidence for criticism of Koyré's claim. In fact, we find that, besides misusing the evidence, Koyré has even changed it. He has done this by smuggling into the Galilean conclusion being considered the qualification "verisimilar," the French *vraisemblable*, which is not in the original. No doubt this is in accordance with his objective of interpreting the qualification as one about the conclusion's empirical truth or correspondence to reality since this is exactly what the term *verisimilitude* suggests. What we find in the original is the qualification "reasonable" in addition to "probable." Translated by Stillman Drake, it reads:

> And therefore if the falling body makes use of the accelerated degrees of speed conforming to the triangle ABC and has passed over a certain space in a certain time, it is indeed *reasonable* and *probable* that by making use of the uniform velocities corresponding to the parallelogram it would pass with uniform motion during the same time through double the space which it passed with the accelerated motion.[8]

Thus insofar as the term *real* in Koyré's claim (G'9) refers to a correspondence to reality, the characteristic being attributed to Galileo's practice is unjustified on the given evidence. But for Koyré, in this context, the term *real* also connotes "temporal" or "pertaining to time," by contrast with the purely spatial. That is, as it is obvious from (G'9), Koyré is claiming that Galileo's constant and sustained attention to the temporal character of the phenomenon of fall led him to his success. But did it?

If one looks in Koyré's account for evidence at all relevant to this aspect of his explanation, one finds two items. The first is Galileo's argument supporting his definition of acceleration. Opening his discussion of naturally accelerated motion in the "Third Day" of the *Discourse,* Galileo writes:

> If we examine the matter ATTENTIVELY, we will find no increase simpler than that which occurs always in the same manner. Now, what this manner is, we will easily comprehend provided we fix our ATTENTION on the *supreme affinity (which exists) between motion and time.* Indeed, as the uniformity and regularity of motion are defined and conceived by the equality of times and distances (in fact, we call uniform a translation in which equal distances are traversed in equal times), so we can conceive the *regularity of the increase in velocity occurring during the same parts of time* by understanding with the mind that

uniformly and, consequently, continuously accelerated motion is that in which *during arbitrary equal times* it acquires equal increments of velocity. [P. 137, Koyré's italics, my capitals]

The second piece of evidence Koyré cites is that Galileo, in addition to giving a proof of the so-called mean-speed theorem gives one for the time-distance law. The mean-speed theorem is the proposition that "the time in which a given distance is traversed with a uniformly accelerated motion by a moving body starting from rest is equal to the time in which the same distance will be traversed by the same body moving with a uniform motion and whose velocity is the mean between the maximum and the minimum for the said uniformly accelerated motion" (pp. 149–50). According to Koyré, "what is lacking to this Galilean demonstration is precisely to show 'the supreme affinity between motion and time,' the preponderant role of time. That is why to this first theorem (the only one which had been proved in the *Dialogue*) the *Discourse* adds a second: 'If a moving body falls from rest with a uniformly accelerated motion, the spaces traversed in arbitrary times are to each other in a ratio double that of the times, that is as the square of the times' " (p. 151).

Koyré's evidence here does indeed indicate that Galileo paid constant and sustained attention to the temporal character of fall. But does it also indicate that this attention enabled him to avoid the error or errors committed by Descartes and formerly by himself? It is difficult to see how that attention could have enabled Galileo to avoid Descartes's error since, as I argued above, Descartes's and Galileo's errors have virtually nothing in common. And Galileo never underestimated the importance of the time-distance law, either at the time of his error or at the time of his success. Indeed it was for him the starting point and raison d'être of the investigations being considered by Koyré. Since Galileo's former error was not that of neglecting the time-distance law, it could hardly have been the attention he paid to time that enabled him later to avoid the error. Hence the present aspect of Koyré's claim (G'9) is not supported by the second piece of evidence.

Is Koyré's first evidential item any more conclusive? Part of Galileo's former error was his assumption of the wrong principle of spatial acceleration. To avoid this error Galileo must avoid this assumption. Attention to time might have led Galileo to the formulation of the correct principle of temporal acceleration, but to account for the

114

rejection of the wrong principle, it is more plausible, as Koyré says elsewhere, "to suppose that his error became apparent to him more directly: by the very fact that his 'axiomatic principle' could not play the role which he wanted to assign to it; it was obviously impossible to deduce from it the descriptive formulas" (p. 107). The correctly deduced formula would have been an exponential function.

But did attention to time really lead Galileo to the successful formulation of the principle? By analogy to the rejection of the wrong principle, is it not more likely that Galileo succeeded because he saw that the principle of temporal acceleration could serve the required function, namely, that the descriptive formulas could be derived from it, thus "demonstrating" them and defining the essence of naturally accelerated motion? True, Galileo tries to justify the principle by appealing to the "supreme affinity between motion and time." He justifies it, but this he does in order to exhibit its simplicity. Attention to the temporal aspect of the phenomenon is important only insofar as it is an instance of the attention to be paid to simplicity. Moreover, the appeal to simplicity is not even the primary aspect of Galileo's justification. For, in the same discussion, he tells us, and the rest of the "Third Day" of the *Discourse* shows, that the correctness of his definition of acceleration is "confirmed *mainly* by the consideration that experimental results are seen to agree with and exactly correspond with the properties which have been, one after another, demonstrated by us." [9]

At any rate it is unclear how any aspect of Galileo's justification practice could account for his success in formulating the correct principle of acceleration. That is, it is impossible, or mysterious at best, that Galileo could have been led to a successful definition of acceleration by the way in which he justified that definition. Rather it must be the way in which he went about trying to find that definition, that is, some aspect of what we may call his investigation practice which is both undeniably present and adequate to produce his result. This aspect, not unknown to Koyré but unappreciated by him (p. 155) is the so-called analytical, regressive, or resolutive method. It consists of Galileo's proceeding from the experimental data, the descriptive formulas, to the principles underlying them. Thus virtually every aspect of Koyré's claim (G'9) must be rejected. Nor is there any other claim constituting Koyré's explanation which is still standing; the explanation has collapsed entirely.

A Modest Alternative

The following alternative explanation emerges from my criticism:

> The march of Galileo's reasoning,—one sees it readily—is faithful to itself. In the *Dialogue* and in the *Discourse* it is the same as in the letter to Paolo Sarpi which we have quoted at the beginning of this study. There, as here, it is, if one may say so, regressive, "resolutive", analytical in the most profound sense of this term. From the fact, from the experimental data, from the "symptoms" of accelerated motion, Galileo ascends—or descends—to its essential definition. There, as here, he looks for the principle, that is to say, the *essence* of this notion, essence, which when translated into acceleration will allow one to deduce and to demonstrate its "accidents" and "symptoms". [P. 155]

Galileo initially chose the wrong principle of spatial acceleration because "it is easier—and more natural—to *see*, that is to say, to *imagine* in space than to *think* in time" (p. 96). His earlier mathematical errors were simply due to his zeal to derive the formulas he knew from the principle as well as to the sheer mathematical difficulty of so deriving them. He then discovered these errors and saw that the principle of spatial acceleration could not serve to derive the formulas. He then considered the more difficult principle of temporal acceleration which was the only obvious alternative. Seeing that he could without fallacy derive the accidents of motion from this principle, he accepted it.

Descartes's method was at no time resolutive. But this is to say, partly, that he faced a greater task, for he was not in clear possession of either the correct descriptive formulas or the correct principle. He was never particularly interested in the problem. When he did give some thought to it, as a result of Beeckman's and Mersenne's questions, his reasoning was altogether confused. It is difficult to see what errors he is making, let alone what their sources are. When he did formulate clear physical principles and consciously adopted a synthetic, or compositive, or deductive approach, he had no particular interest in the problem, especially because of the complexity of its solution.

To summarize: Galileo's success is due to his resolutive methodological practice, his initial error to a choice of an easier principle. Descartes's error remains a mystery.

chapter 7

Explaining the Rise of Modern Science

THE EXPLANATION OF the rise of modern science is an activity whose problems, unlike those of Guerlac's and Koyré's explanations, are well known. My discussion of them will thus not need the extensive documentation and analysis required in the two previous chapters. A critique of recent literature on the subject is necessary, however, to focus attention on the pertinent issues.

I shall first state the conflicting explanations of the rise of modern science and the dissatisfaction felt by historians themselves with the explanations. My critique will show that one is faced with a confusion rather than a conflict of opinion because there are important similarities between presumably different views and important differences between presumably similar ones. I shall then examine critically a leading historian's attempt to sort out the confusions. Finally, I shall construct a simple logical analysis based on the concepts prevailing in the controversy—sociology of science, intellectual history of science, externalism, and internalism—showing that they cannot provide a satisfactory framework for explaining the rise of modern science.

The Conflict of Explanations

The conflicting explanations of the rise of modern science are well illustrated, though not so well discussed, in *The Rise of Modern Science: External or Internal Factors* (1968), a volume in the "Problems in European Civilization" series.[1] The editor, George Basalla, sums up the problem, as he sees it, in his introduction:

> Modern science is a product of Western civilization. It first appeared in Europe during the sixteenth and seventeenth centuries and then slowly spread to North and South America, Australia, Asia, and Africa. In this way the fruits of modern science were made more widely available to the world. These fruits included novel and fundamental insights into the operations of nature, a high-level of material well-being, and, most recently, the possible destruction of all life on earth.
>
> How did this vast enterprise get its start? What were the unique elements in the Western tradition that stimulated its creation and rapid growth? Did science emerge because of a mutation in the intellectual life of Europe or through a long developmental process? Does its origin and growth depend upon external factors—e.g. social and economic conditions—or upon factors within science itself? These questions have perplexed historians for over a century. In attempting to answer them historians have turned to the histories of science, economics, religion, intellectualism, psychoanalysis, political ideology, art and the occult, and sociology.[2]

After briefly reviewing the articles included in his collection, Basalla documents the conflict of opinion:

> The traditional explanation stressed the sterility of the Middle Ages and the sudden appearance of modern science in the Renaissance:
> "We have now to consider more especially a long and barren period, which intervened between the scientific activity of ancient Greece and that of modern Europe; and which we may therefore, call the Stationary Period of Science. . . ."
> —William Whewell

To which the medievalists replied:

"The mechanical and physical science of which modern times are justly proud unfolds, through an uninterrupted series of barely imperceptible improvements, from the doctrines taught in the medieval schools."

—Pierre Duhem

And the Renaissance scholars countered:

"When Duhem, the arch-opponent of Burckhardt, reached a vantage point from which he could look back over his beloved fourteenth century and forward to the sixteenth and seventeenth centuries, to all intents and purposes he withdrew his objections to Burckhardt's thesis of a Renaissance in science and gave it his valuable though grudging support."

—Edward Rosen

Various explanations were offered to account for the rise of modern science in the sixteenth and seventeenth centuries:

"Science flourished step by step with the development and flourishing of the bourgeoisie."

—Boris M. Hessen

"A number of studies have shown that the Protestant ethos exerted a stimulative effect upon capitalism. Since science and technology play such dominant roles in modern capitalist culture, it is possible that tangible relationships exist between the development of science and Puritanism."

—Robert K. Merton

"Ancient science failed to develop not because of its own immanent shortcomings but because those who did scientific work did not see themselves, nor were they seen by others, as scientists, but primarily as philosophers, medical practitioners, and astrologers. . . ."

—Joseph Ben-David

"The scientist of the seventeenth century was a philosophical optimist; delight and joy in man's status pervaded his theory of knowledge and of the universe. And it was this revolution in man's emotions which was the basis for the change in his ideas."

—Lewis S. Feuer

". . . this investigation seems to suggest that two of the major features of the Scientific Renaissance, namely the change-over from Form to Function, and the rise of a 'natural law' unconnected with the affairs of human society have their origin in a specific transformation in the arts. . . ."

—Giorgio De Santillana

"One cannot deny that a careful analysis of the alchemical texts, pharmaceutical works and the metallurgical treatises of the Renaissance for their actual chemical content is of profound importance for our knowledge of the growth of science as we know it, but the blanket dismissal of other supposedly 'non-scientific' aspects of early chemistry to the realm of occultism, mysticism, and magical hocus-pocus does nothing to add to our knowledge of the birth of modern science."
—Allen G. Debus

But all explanations based on external factors have been challenged by the internalists:

"Clearly, externalist explanations of the history of science have lost their interest as well as their interpretative capacity. One reason for this may be that such explanations tell us very little about science itself. . . . Social and economic relations are rather concerned with the scientific movement than with science as a system of knowledge of nature (theoretical and practical); they help us to understand the public face of science and the public reaction to scientists; to evaluate the propaganda that scientists distribute among themselves, and occasionally—but only occasionally—to see why the subject of science takes a new turn."
—A. Rupert Hall [3]

Conflict of opinion is not necessarily a sign of an unsatisfactory state of affairs. But this conflict is so characterized by historians of science themselves. For example, Thomas Kuhn, enough of an authority and spokesman for the profession to have written the article on "History of Science" in the entry "Science" in the *International Encyclopedia for the Social Sciences*,[4] thinks that conflict of opinion is an unsatisfactory state because it is a sign of the "pre-scientific" state of research when the given practice has not reached "maturity." He regrets that:

there seem at times to be two distinct sorts of history of science, occasionally appearing between the same cover but rarely making firm or fruitful contact. The still dominant form, often called the "internal approach", is concerned with the substance of science as knowledge. Its newer rival, often called the "external approach", is concerned with the activity of scientists as a social group within a larger culture. *Putting the two together is perhaps the greatest challenge now faced by the profession,* and there are increasing signs of a response. Nevertheless any survey of the field's present state must unfortunately still treat the two as virtually separate enterprises.[5]

However, what we actually have is not, or not merely, a conflict of opinion about the rise of modern science but primarily a confusion of opinion. Suppose we adopt an empirical approach and interpret the conflict to be the one between historians like A. Koyré, H. Butterfield, A. R. Hall, A. C. Crombie, Marshall Claggett, E. A. Moody, J. H. Randall, and E. A. Burtt on the one hand, and "sociologists" like B. M. Hessen, R. K. Merton, E. Zilsel, and G. N. Clark on the other. A survey of the relevant literature shows this division to be the one most readily mentioned; it is the one explicitly suggested, in part, by both Basalla[6] and Hall.[7] This division, however, is only apparent; in reality, we have a confusion.

First, suppose we examine the relevant works of Koyré, the leading internalist, and of the Marxist Hessen, the most extreme and uncompromising externalist, about whom Basalla claims "there is general agreement that [he] distorted historical facts to fit his ideological mold."[8]

In Hessen's essay "The Social and Economic Roots of Newton's Principia," one of his major claims is that "Newton's appeal to a divine mind as the highest element, creator, and prime motive power of the universe, is not in the least accidental, but is the essential consequence of his conception of the principles of mechanics,"[9] and more generally that "the idealistic views of Newton are not accidental but organically bound up with his conception of the universe."[10] While one of Koyré's major theses in his essay "Newton and Descartes," is that Newton's "preoccupation with philosophical problems was not an external *additamentum* but an integral part of his thinking."[11] Not only do Hessen and Koyré agree on this central thesis, but they give analogous arguments to support it; their arguments, in fact, consist of an analysis of the content of some of Newton's writings. Koyré, of course, uses an additional document, the recently discovered and published early paper by Newton, *De gravitatione et aequipondio fluidorum,* but this does not change the general form of his argument.

Moreover, we find G. N. Clark, a "sociologist of science," giving what is in effect a defense of the importance of the primary motive used by the "intellectual historian":

> The disinterested desire to know, the impulse of the mind to exercise itself methodically and without any practical purpose, is an independent and unique motive. We might examine it by

121

tracing through the thought of the sixteenth and seventeenth century the distinction between pure and applied science [which is what the intellectual historians of science might do]; but the quickest and clearest way of disentangling it is to abandon the biographical point of view and take that of history. . . . [To do this means realizing that] the scientific movement was not the mere aggregate of the efforts of many individuals, each one of whom can be explained in terms of his social position. It was a common enterprise, partially incarnated in each of them, but having its own existence and nature, not to be explained except as a whole greater than its parts. . . . [For example] the pursuit of knowledge in the universities is a self-perpetuating tradition . . . [so that] it was the social function of the universities thus to set free from the pressure of other motives men who had the desire to know. The changing social and economic conditions of Newton's time did something to increase this social provision, especially for mathematics and science; but the universities had their own laws of growth. At their heart was the disinterested love of truth. This we must add as a sixth, and greatest, to the five motives which we have distinguished as actuating the scientific movement [i.e., the influences "from economic life, from war, from medicine, from the arts, and from religion"]. The others helped to clear the way for it, they could not and did not create it.[12]

Finally, it is well known that the intellectual historians are divided among themselves on the question of the connection between seventeenth century science and medieval science. Koyré emphasizes the differences between the two, whereas most of the other intellectual historians emphasize various kinds of similarities. And this internal conflict among intellectual historians is rather basic since it centers around the question of whether, as a matter of principle, one should look for similarities or for differences in historical developments. In the fundamental idea of contrasting seventeenth century and medieval science, most externalists would agree with Koyré.

Hall's Analysis

What are we to make of the conflict between the so-called internalists and externalists when we find such important cross-group similarities and intra-group differences? A. R. Hall attempts to sort out the supposed issues between them in his article "Merton Revisited, or Science and Society in the Seventeenth Century": [13]

one issue between the externalist and internalist interpretation
is this: was the beginning of modern science the outstanding
feature of early modern civilization, or must it yield in importance to others, such as the Reformation or the development of
capitalism? [14]

But this cannot, or at least should not, be one of the main issues since,
if anything is clear, it is that the dispute is one about the causes of the
rise of modern science, not about the importance of a historical development, which is a dispute about effects rather than causes. In fact, the
relative importance of various historical developments should be judged
by their independent effects. Given two events E and E', the effects of
E independent of E' are those events which are effects of E but which
could not have been effects of E' unless E' were itself an effect of E.
Therefore, in judging the relative importance of the rise of modern science, the rise of capitalism, and the Protestant Reformation, even if the
former had been caused by either or both of the latter, so that indirectly
all the effects of the rise of modern science would also be effects of the
rise of capitalism and/or the Reformation, even so the rise of modern
science would be the more important event. The reason for this is that
the important determining effects, such as the material progress of
modern Western civilization, the spread of science throughout the
earth, and the threat of the bomb, are consequences of the rise of
modern science irrespective of what its causes were.

Hall also regards as an issue the nature of the connection between
religion and science in the seventeenth century. He believes that the ex-
ternalist-sociologists would affirm, and the internalist-intellectual historians deny that the connection, besides being a correlation, is causal.
But who, intellectual historian or not, has ever denied or could deny
that the Inquisition had an effect on science in Italy and France? Can
there be any doubt that to burn a Bruno, to imprison a Galileo, or to
discourage a Descartes from publishing *Le Monde* is to have a causal
influence on science? The influence here is detrimental to be sure, but it
is nonetheless causal. As for the more restricted question of the connection between Puritanism and science in seventeenth century England, it is true that Merton does not draw the conclusion about their
causal connection. But this is probably due to his being really interested in sociological-theoretical conclusions and no such conclusions
really follow from the historical material he presents. But if it is a
historical thesis that we are interested in, that is if we are interested

in the claim that in seventeenth century England it was the rise of the Puritan ethic which caused the increase of interest in science, then it is not clear why "massive proof and argument is required in demonstration," [15] as Hall says. Why is it not sufficient to do what is often done in history in order to establish a causal claim,[16] namely, to establish that:

(1) the event E occurred,
(2) condition C could have caused E,
(3) nothing else was present which could have caused E, and that
(4) therefore (since every event must have a cause) C must have caused E.

It is true that Merton has established that $(1')$ there was an increase of interest in science in the latter part of the seventeenth century in England, and $(2')$ that the rise of the Puritan ethic could have caused such a development. And this latter thesis is nontrivial especially in view of the usual beliefs about the enmity between science and religion. But Merton has not argued in support of $(3')$ the idea that none of the other conditions present in England at the time could have caused the increase of interest in science. It is for this reason that the causal conclusion cannot be drawn from Merton's work. To question the need of (3), in addition to (1) and (2), for the inference of (4) would raise a genuine issue, but that would not allow us to make the desired historiographical distinction; in fact, Merton, an externalist-sociologist, does not, in the absence of the analog of (3), i.e., in the absence of the elimination of other possible causes, infer the causal claim; whereas Koyré in his "Newton and Descartes" does draw the causal conclusion from the possibility claim that Newton's philosophical anti-Cartesianism could have caused him to write the *Principia*.[17]

Is the issue that of distinguishing properly between seventeenth-century science and technology? Hall suggests this when he criticizes Merton on the grounds that "an analysis that confuses mathematical physics with mathematical technology in this way bewilders rather than assists the historian of science." [18] But this cannot be one of the main issues between the two schools for the following reason. The externalists-sociologists actually do distinguish between technology and science; they make the distinction when they put forth their externalist explanations, for those explanations could not even seem externalist unless technology were being regarded as something different from science. It is quite another matter, of course, to say that the externalists-sociologists have shown that there was a causal influence of technology on science in the seventeenth century. Hessen and Zilsel would be the only ones who would claim that they have established, or are trying to establish, a causal claim: Hessen, that Newton wrote the *Principia* in order to provide a systematic and theoretical solution to the technological problems of his age; Zilsel, that the rise of modern science was caused by the transposition of the methods of technology (arts and crafts) to natural philosophy. I think that both have succeeded in establishing the weaker possibility claim that the given event could have been caused by the correlative factor. And only in the absence of other possibly causal factors can one assert the actual causal claim. In the case of Newton's *Principia* it is obvious that other determining reasons (e.g., the one adduced by Koyré) were present; whereas I am inclined to think that other than those factors adduced by Zilsel, none were present which would account for the pioneering work of Gilbert, Bacon, and Galileo.

From what I have said above, possibility causal claims may seem easy to establish. This is not so; e.g., no one has been able to establish that the rise of capitalism could have had a causal influence on seventeenth-century science. It may appear that this is what Hessen has established; but that is only what he wanted to establish. He has not established that the rise of capitalism qua capitalism, i.e., qua economic system, could have determined Newton's work. He succeeds in showing only that capitalism, qua technology, may have determined Newton's work. That is, the technological problems would have been present, and could have caused Newton's work, even in a system of socialism.

Is one of the main issues the question of how the large-scale outline of the science of a given epoch is to be accounted for, as Hall also suggests? [19] Here the externalists-sociologists would presumably say that it is to be explained on the basis of external social factors influencing the science; whereas the internalists-intellectual historians would claim that the large-scale outline, as well as the smaller and more specific details, should be accounted for in terms of the internal intellectual state of the science. But this cannot really be an issue since, *pace* Clark, the gross characteristics of the science of a given epoch are the sum-total of the small-scale characteristics, and, if these latter are determined by the internal state of the science, so are the former. Only by changing the kind of description when we move from small-scale to large-scale characteristics, which amounts to changing one's point of view, can we change the kind of determining factor. For example, by changing from an intellectual description of a small-scale development to an institutional description of a large-scale development can one change the kind of explanatory factor. But the change is due to the change from intellectual to institutional considerations and not to the change from small to large scale. Through this sort of change-of-description subterfuge most externalists-sociologists bring in their social "factors." But there is no good reason why nonintellectual descriptions of large-scale developments should be considered, or, if they are considered, why one should eliminate nonintellectual descriptions of small-scale developments, which would then have to be explained by "external" factors. To cite an example: an intellectual description of a large-scale development would be "the transition from the geostatic to the heliostatic astronomical theory"; a description of a small-scale nonintellectual development would be "the trial of Galileo." The former has to be accounted for internalistically-intellectually, the latter sociologically. Thus even if we abandon the "heroic view of history," as Merton urges us to do, or the "method of biography," as Clark calls it, we need not necessarily take the point of view of "externalism-sociology of science." The issue under consideration therefore cannot represent a meaningful disagreement between the two schools since it originates from different initial points of view and thus reflects different interests. Hall's suggestions render the conflict of explanations of the rise of modern science no less confused. He is right, however, in concluding that the resolution of the conflict-confusion "will require a fine analysis, a scrupulous drawing of distinctions." [20]

Unproductiveness of One Conceptual Analysis

So far I have followed an empirical approach in beginning with a widely accepted empirical division of historians of science into two groups and then showed that no such division was justifiable. Second, I located a set of claims believed by many historians to define certain differences and showed that the differences so defined are not valid. I shall now apply a more conceptual method to the problem.

Let us consider what the real differences are between what might properly be called "sociology of science" and what might properly be called "intellectual history of science." Intellectual history of science should concern itself with the intellectual developments constituting the history of science. The sociology of science should concern itself with events, phenomena, and developments involving social institutions and large numbers of individuals. Hence there is no conflict between the two: an intellectual development might involve directly and primarily one individual (e.g., Newton's discovery of the law of gravitation) or a large number of individuals (e.g., the rise of modern science); and a development involving many individuals may be either intellectual (e.g., the transition from geostatic to heliostatic astronomy) or not (e.g., the increase in the relative number of scientists in the latter part of the seventeenth century in England). Consequently, a social-intellectual history of science is a real possibility, and there ought to be no historiographical disagreement between the sociologists and the intellectual historians. If there is a disagreement, and it is a real one, then we have hereby been unable to discern it.

On the other hand, if we use the concepts of internalism and externalism to define the difference, then the issue becomes ambiguous and inapplicable to the historiographical situation. The ambiguity is due to the different things with respect to which internalism and externalism can be internal and external. For example, since scientific thinking cannot be equated with all of intellectual activity, an intellectual internalist approach could be externalist with respect to developments in scientific thinking. The question could arise whether modern scientific ideas originated from previous scientific ideas or from previous philosophical ideas. This question acquires interest and importance in Kuhn's theory of scientific change, but it has not been explicitly or frequently discussed in the recent controversies. Koyré would turn out to be an externalist on such an issue. Things that could be regarded as external to all ideas, scientific or nonscientific, are facts and practices. In some obvious sense such things are more external to (unlike) scientific ideas than philosophical or metaphysical ideas are. Yet are not *scientific* practices more internal to science than metaphysical *ideas?* Whatever the case, Koyré's work, once more, would have externalist tendencies. I am referring to the possibility of ideas originating from certain procedures. For example, Galileo's law of falling bodies is alleged by Koyré to result from certain practices. It may be objected that these practices are, however, intellectual or mental practices. But then we are distinguishing between intellectual ideas and intellectual practices. And the question arises whether in-

127

tellectual practices are more like (internal to) other kinds of practices or more like intellectual things of other kinds. Would religion, economy, and technology be external to science if one dealt with religious, economic, and technological ideas? The externalism-internalism distinction does not seem too helpful. Thus the fine analysis and scrupulous drawing of distinctions that Hall urges, cannot be the present ones.

PART TWO

Philosophy of the History of Science

chapter **8**

Inductivism, Conventionalism, and Explanationism

SO FAR I have argued that the concept of history-of-science explanation is anomalous from the point of view of predictiveness, law-coverability, and the 'why' principle. I have also argued that the practice of explanation is unsatisfactory in the sense that the most obvious characteristic of historical accounts which are *explicitly* explanatory is that they are highly unsatisfactory. Several questions come to mind as a result of this anomaly and deficiency: Are these two things connected; if so, how; and, how is one to understand the causes of this unsatisfactory situation? Such queries constitute what I shall call "the problem of explanation in historiography of science." The first step toward its solution will be an examination of previous attempts to solve it. The single investigation most relevant to the problem so far is that undertaken by Joseph Agassi in *Towards an Historiography of Science*.

Agassi's most pertinent thesis is his claim that "historical explanation of any value is rare in the annals of the history of science, mainly because of a naive acceptance of untenable philosophical principles" (pp. viii, 74–79). Since he is rather brief about this claim, and since this claim is a special case of his main thesis, I shall begin by examining his main thesis, which he summarizes:

> The history of science is a most rational and fascinating story; yet the study of the history of science is in a lamentable state: the literature of the field is often pseudo-scholarly and largely unreadable. The faults which have given rise to this situation, I shall argue, stem from the uncritical acceptance, on the part of historians of science, of two incorrect philosophies of science. These are, on the one hand, the *inductive philosophy* of science, according to which scientific theories emerge from facts, and, on the other hand, the *conventionalist philosophy* of science, according to which scientific theories are mathematical pigeon-holes for classifying facts. The second, although some improvement over the first remains unsatisfactory. A third, contemporary theory of science, Popper's *critical philosophy* of science, provides a possible remedy. On this view, scientific theories explain known facts and are refutable by new facts. [P. v]

The Inductivist Interpretation

In order to extract a coherent view from Agassi's essay, his main thesis may be analyzed into four parts: the first is an evaluative claim, the second is a causal one, and the last two are methodological claims. The evaluative claim is the assertion that "the study of the history of science is in a lamentable state" (p. v). The causal claim is the assertion that "the faults which have given rise to this situation, I shall argue, stem from the uncritical acceptance, on the part of historians of science, of two incorrect philosophies of science" (p. v). The first methodological claim is the general prescription that the historian of science ought to be (more) critical about the philosophy of science he accepts (pp. 14, 29, 45). The second methodological claim is the specific proposal that the historian of science ought to try to apply Popper's "critical" philosophy of science to the history of science (pp. v, 78).

Perhaps the most obvious relation among the first three claims is that they form an argument of which the methodological claim would be the conclusion. This form of argument is popular among methodologists, and may be regarded as "the" methodological argument. Its structure is:

Condition E is bad (in some respect).
The cause of E is condition C.
Therefore, eliminate or try to prevent the occurrence of C.

Agassi is thus grounding his general methodological claim on his evaluative and causal ones. Hence the question of whether or not to accept the general prescription partially reduces to the question of his evaluative and causal claims' acceptability.

But how does Agassi support his specific methodological claim that the historian of science ought to apply Popper's philosophy of science to the history of science? One way is by arguing that it is a special case of the general claim. In fact, Popper's philosophy of science, so the argument could go, is a relatively recent philosophy. Because it hardly has been applied to the history of science, the historian of science will necessarily accept it self-consciously. This does not mean that he would be adopting it "on faith"; for philosophically it easily can be seen to be superior to inductivism or conventionalism. On the other hand, the willingness to test Popper's philosophy of science would be identical with the attempt to apply it, which is what the prescription under consideration enjoins one to do. Thus for the historian of science to try to apply Popper's philosophy of science is to be critical about the philosophy of science he accepts, and is what the general methodological claim prescribes. Thus the acceptability of both of Agassi's methodological claims depends on the acceptability of the evaluative and causal ones.

Let us see therefore whether these claims are acceptable. The evaluative claim is grounded on a series of specific evaluations which fall into five main categories. Two of these are mentioned in the "Introductory Note" (p. v): the literature in the field is (a) often pseudoscholarly and (b) largely unreadable. The other categories are what Agassi calls (c) the "cancerous growth of continuity" (pp. 33–40), (d) "wisdom after the event," and (e) the rarity of historical explanations of any value (p. viii, and pp. 74–79).

If the main causal claim is to be tenable, the uncritical acceptance of inductivism or conventionalism must be the operative cause in each of the cases falling under each of the above five categories. A full examination of Agassi's thesis would then require the examination of such a series of pairs of evaluative and causal claims. We shall not undertake such a full examination, for it must be admitted that to reconstruct Agassi's general argument in the manner indicated is to make him look like an inductivist. For Agassi, the inductive philosophy of science claims that scientific theories emerge from the facts (p. v); and more generally the inductivist philosophy claims that theories or ideas emerge from facts. It is not clear whether, in speaking of "emergence," Agassi refers to logical grounding or to historical-psychological emergence; but, whichever he means, the above reconstruction of his argument would make his thesis "emerge" from the facts of the historiography of science. And in order to give a fair characterization of his argument one should not portray him as an inductivist, however natural such a portrayal may seem, as it is the case in the above reconstruction. Of course the possibility remains that Agassi is being inconsistent, but one should try to avoid interpretations which attribute obvious inconsistencies to an author.

The Conventionalist Interpretation

The curious thing is that we can also find grounds for a conventionalist interpretation of his thesis and argument. According to him,

> the central doctrine of conventionalism is that scientific theories are neither true nor false, that their general frameworks are mathematical systems which serve as pigeon-holes within which to store empirical information. Which pigeon-hole system to adopt is a question of choice for which simplicity provides the criterion. We can rearrange a pigeon-hole system or change it without thereby proving its falsehood, or unscientific character, or badness. Theories can fit facts with greater or smaller degrees of simplicity. Hence simplicity is a criterion not of absolute, but only of relative merit; it is a substitute for the absolute criterion of merit of the inductivist. [P. 29]

The first trace of conventionalism is found in the "Introductory Note" itself, where Agassi claims that both the inductivist and the conventionalist philosophies are incorrect, not false but "incorrect." Second, he claims that though they are both incorrect, the conventionalist philosophy is better than, "some improvement over," the inductivist philosophy. Third, he claims both in the note and in the last section (sec. 18, p. viii, pp. 74–79) that the main fault in the historiography of science is not the incorrectness or untenability of those philosophies, but the uncritical or naive acceptance of them. Fourth, we are told in section 8 not that the "conventionalist view" has led to important and interesting historical discoveries, but that it "has proved a *useful tool* in the hand of the historian of science" (p. 29, my italics). Finally, in the descriptive title of section 12, conventionalism is called a "framework" (p. viii). This verbal evidence merely suggests Agassi's conventionalism. That his main argument can be given a conventionalist interpretation shows that his practice also is conventionalist.

This argument maintains that the facts of the development of science can be classified in various ways, of which one is to classify them by their relation to the latest science textbook. Thus ideas are classified as scientific or unscientific depending on whether or not they are found in the up-to-date textbook; individuals, all and only those whose ideas agree with the textbook are scientists. Finally, events and developments are regarded as being part of the history of science, i.e., as "scientific," if and only if they have left a trace in the textbook; otherwise they are part of the history of magic, superstition, error, metaphysics, philosophy, etc. Thus "the simplest formula for an inductivist history of science is to arrange the up-to-date science textbook in chronological order, to describe some of the circumstances surrounding the occurrence of an important event in the history of science, and say something about the chief actors involved in that event" (p. 2).

The inductivist philosophy of science is indeed the simplest way of classifying the facts of the development of science when little is known about the causal relations between scientific and nonscientific or "unscientific" events, ideas, and thinkers. And such was the case during what Agassi calls the "golden age of inductivism," approximately from Bacon to Duhem. That is why the work of those historians of science is, in general, unobjectionable. Not even the "black-and-white categorizing," as Agassi calls it, is objectionable, for it is a necessary consequence of the criterion of classification: the item to be classified either is or is not related to the textbook and there is no third alternative.

Inductivism is unhelpful, however, in classifying facts involving connections between black and white. Some such facts were known even during the golden age of inductivism (p. 30); but as long as there were not too many, they could be simply ignored. But "Duhem has shown that such cases are numerous" (p. 31). The simplest way of making the classification is now to use a relative criterion, to replace the absolute one of the textbook: the simplicity of the idea or theory is such a criterion. Thus a history of science is no longer an account of the temporal evolution of the latest textbook, but an account of the evolution of simpler and simpler theories for the classification of the facts known at the time (p. 30).

Conventionalism remains the simplest framework as long as one ignores either the existence or importance of struggles, conflicts, controversies, and dissent. That is, in fact, what the conventionalists do. For example, Whittaker, a leading conventionalist, "has entirely ignored the conflict between classical determinism and modern indeterminism in physics" (p. 25). Moreover, "nowhere in the literature on the history of physics [has Agassi] found discussed the fact that methodological disagreements have been as common in science as in philosophy" (p. 25). There is also the fact that "Priestley's dissent from the French school of chemistry is historically important, yet it does not fit the conventionalist framework because conventionalism too leaves little room for controversy" (p. viii, p. 45–48). As another example, the conventionalist "ignores the fact that, historically, Lavoisier's behavior can be understood as an attempt to examine critically the doctrines of Stahl and Black" (p. 50); he also ignores "the logical conflict between Aristotle and Archimedes in the history of Renaissance mechanics" (p. 57) or that between "Galileo's terrestrial mechanics and Kepler's celestial model" (p. 58). Finally, though Duhem describes Oersted's discovery of electromagnetism as " 'the point of departure of all researches which constitute electrodynamics and electromagnetics' " (p. 67), he ignores the fact that "Oersted's discovery . . . conflicted with Newton's theory of force" (p. viii).

But, continues Agassi, such facts are too numerous and important to be ignored, especially now that Koyré has promoted studies for the discovery of such facts (pp. 24, 57). And it is no accident that the conventionalist ignores such facts; for conventionalism is an *in toto*, *post mortem* (p. 48) criterion. In saying that it is an *in toto* criterion, Agassi means, I think, that it is a criterion for the classification of large scale facts, that is, facts involving wholes, collectives, or "communities" of scientists; in other words, it is a collectivist or holist criterion. And this criterion, like all such criteria, "misses a main driving force of science, the *thinker's* problems and his sleepless nights" (p. 41, my italics), and the "intellectual obstinacy, dissatisfaction, so characteristic of scientific research, even when conducted by a calm, well-adjusted and placid person" (p. 48).

In saying that the conventionalist criterion of simplicity is *post mortem*, Agassi means that it "is *useless* when it comes to reconstructing the struggle between schools, and weighing the merit of the criticisms levelled by each school against each other" (p. 49, my italics). The reason is that, "assuming that sooner or later the simplicity of one of the competing doctrines becomes obviously greater than that of the other, the conventionalist has to dismiss one school even before it ceases to be scientifically important and interesting" (p. 45).

What criterion shall we then choose for the classification of the facts of the history of science? As a good conventionalist in the field of the history of science, Agassi realizes that the reason why "the conventionalist view that we prefer the simple theory to the less simple one, not a true theory to a false one, has proved a useful tool in the hand of the historian of science . . . is not . . . the admitted importance of simplicity [i.e., not the, perhaps approximate, *truth* of the conventionalist view, as an inductivist might say], but the introduction of a new criterion of graded valuation to replace the old inductivist criterion which divides theories into the good and the bad" (p. 29). Thus the new criterion that we need now, to replace simplicity, is one which retains the continuum of evaluations but allows one to classify the known and soon to be discovered facts about controversies; the new criterion will have to admit both that scientific theories may be false and that they are criticizable (p. 49). Explanatory power is such a criterion. On this view, scientific theories are supposed to explain known facts and are criticizable by new facts (p. v). Under this new classification of the facts of the history of science, the history of science becomes the never ending struggle among the various schools of scientific thought (p. 23) for the attainment of a true explanatory theory, the achievement of which is regarded as the aim of science.[1]

The preceding argument would support two conclusions. First it would substantiate the Popperian philosophy of science, whose central doctrine is that scientific theories are explanations of facts, on the grounds that this philosophy is the simplest classification of the facts of the development of science. More generally, it would support the view that philosophies of science, that is, views like inductivism, conventionalism, and Popperianism, are neither true nor false but classifications, more or less adequate or "simple" classifications, but classificatory frameworks nonetheless, of the facts of the development of science. According to this conventionalist interpretation we would then be attributing to Agassi a Popperianism in the field of the philosophy of science, i.e., for him, in the domain of the facts of the development of science; we would also be attributing to Agassi conventionalism in the field of the philosophy of the historiography of science, i.e., for him, in the domain of the facts of the development of the historiography of science. But it is clear that Agassi is not a conventionalist, least of all in the domain of facts of the development of the historiography of science. For example, he gives a Popperian interpretation of the rise of the conventionalist historiography of science: not only is the conventionalist philosophy of science made to emerge prior to conventionalist historiography, but also he explicitly refuses to apply the "emergence technique" when giving an account of the origin of Duhem's philosophy; Duhem is alleged to have developed his philosophy, primarily, though not exclusively, in order to criticize inductivism (pp. 32–33). And, in general, it is Agassi's "frank intention to advocate Popper's methodology as a means of improving the present lamentable state of affairs in the field of the history of science" (p. 78), that is, to apply Popper's philosophy to the historiography of science.

The Popperian Interpretation of Agassi's Essay

It seems then that a fair interpretation of Agassi's main thesis and argument can neither ground that thesis on the facts of the historiography of science, by starting with those facts and making the thesis "emerge" from them, nor by arguing that the simplest classification of the facts of the development of science is now the Popperian philosophy of science. This does not mean, however, that our two previous efforts for inductivist and conventionalist interpretation of Agassi will have been wasted. For the possibility of such interpretations, which should be beyond dispute from the above arguments, is significant and can be used to acquire some insight into inductivism, conventionalism, and Popperianism (hypothetico-deductivism).

The Popperian interpretation obviously will have to attribute to Agassi, as an important element of his thesis, Popper's so-called "critical" philosophy of science, the idea that scientific theories are explanations of facts. And the argument in favor of this philosophy will have to be that it is the only explanation, or at any rate the only explanation of many or some of the important facts of the development of science. Hence another element of Agassi's thesis is the idea that philosophies of science are explanations of the facts of the development of science. But what does this idea allow us to explain? Let us remember that, in speaking of philosophies of science, Agassi is here referring primarily to inductivism, conventionalism, and "Popper's critical philosophy of science"; but then it is obvious that he is using the idea just mentioned in his interpretation of the development of the historiography of science. Thus that idea, which may be regarded as the central tenet of Agassi's philosophy of the historiography of science, is supposed to explain the facts of the development of the historiography of science.

We now see why the Popperian "critical philosophy of science" is not and cannot be, for Agassi, the "panacea"; for the panacea is Agassi's own philosophy of the historiography of science. And we also see why "Popper's critical philosophy of science" is just "a possible remedy." For, as we have already seen, what both the inductivists and conventionalists are urged to do is not necessarily to accept Popper's philosophy of science, and thus abandon their own philosophies of science (only an inductivist or conventionalist philosopher of the historiography of science would advocate such a change). Rather they are being urged to accept Agassi's philosophy of the historiography of science, which in fact provides a justification why they should continue to be either inductivists or conventionalists in the domain of the facts of the development of science, provided they are willing to articulate their own philosophies of science which now happen to have less explanatory power than Popper's philosophy of science. Of course the latter alleged fact would be, for someone without prior commitment or other inclination, a sufficient reason to accept Popper's critical philosophy of science.

But why should the inductivist or conventionalist historian of science, or in fact anybody else, accept Agassi's philosophy of the historiography of science? Agassi's argument, as we have seen, would have to be that this philosophy has great explanatory power, perhaps more than any of its alternatives. The alternatives would be what might be called the inductivist and the conventionalist philosophies of the historiography of science. But, this being the case, Agassi's argument, even if valid, could not, and should not by his own criteria, convince someone who had prior commitment or other inclination to accept either one of the two alternative philosophies mentioned. But such people are usually, and naturally enough, the inductivist and conventionalist historians of science themselves. For they, like Agassi, as he himself notes in the case of inductivism (p. 1), do have the implicit and explicit inclination to extend the inductive and conventionalist philosophies of science, which they perhaps uncritically accept, first from the domain of natural phenomena into the domain of facts of the development of science, and then into the domain of facts of the development of the historiography of science. Most likely, then, they accept either the inductivist philosophy of the historiography of science, according to which philosophies of science emerge from the facts of the development of science, or the conventionalist philosophy of the historiography of science, according to which philosophies of science are frameworks for the classification of facts of the development of science. Thus inductivist and conventionalist historians of science cannot be forced to accept Agassi's philosophy of the historiography of science by the argument that this philosophy has more explanatory power than either of its alternatives.

Thus, to convince inductivists and conventionalists may be the, perhaps unconscious, reason why there are traces, and indeed more than traces, of both inductivist and conventionalist arguments in support of Agassi's philosophy of the historiography of science. For the inductivist argument tries to show that, by extending inductivism into the domain of facts of the development of the historiography of science, there "emerges" Agassi's philosophy of the historiography of science and, as a conditional consequence, "Popper's critical philosophy of science." And the conventionalist argument tries to show that, by extending conventionalism into the domain of the facts of the development of science, the simplest classification of these facts is the Popperian critical philosophy of science. And once we have the Popperian philosophy of science, it is obvious that by extending it into that same domain we get Agassi's philosophy of the historiography of science. Thus, if the two arguments mentioned in our two previous interpretations of Agassi are sound, even the inductivist and conventionalist have to accept Agassi's philosophy of the historiography of science, though not necessarily Popper's critical philosophy of science.

But my aim is not to preach Agassi's philosophy of the historiography of science to the inductivist and conventionalist historians; indeed I do not think that those arguments are sound. Therefore I shall not give an explicit analysis and evaluation of the two previous reconstructions; but since I wish to criticize Agassi's thesis, I shall concentrate on the Popperian interpretation of Agassi's thesis and argument. The point of view that I shall take is that of someone who accepts Popper's philosophy of science. The evaluation of Agassi's essay then reduces to the determination of whether someone who accepts Popper's philosophy of science should accept Agassi's philosophy of the historiography of science.

My own views inductivistically emerged from this critique of Agassi. First, Agassi's argument in favor of his philosophy of the historiography of science is without force because all of his explanations of facts of the development of the historiography of science are highly unsatisfactory by his own standards, which in this case are the so-called deductive model of explanation. Second, the views to which he is referring in his statement of his philosophy of the historiography of science, namely, inductivism, conventionalism, and "Popper's critical philosophy of science," are not explanations of the facts of the development of science; rather, inductivism is the common method of exposition, presentation, or justification; conventionalism is the externalist point of view about science; and the hypothetico-deductivism of Popper's philosophy of science is a common method of discovery; a partial justification of this claim can be found in my solution to the problem of explanation. Third, what Agassi really means by "philosophies of science," namely, "philosophies of the history of science" or "theoretical histories of science," are not explanations (in Agassi's sense) of the facts of the development of science, but "emerge" from those facts (this may be called my "inductivist philosophy of the historiography of science"). Agassi's argument in favor of Popper's philosophy of science is also without force because his explanations of the facts thereby explained would be highly unsatisfactory, especially by his own standards; I shall not explicitly argue for this, but my argument in chapter 12 against philosophies of the history of science will provide a partial justification. Finally, though I do not accept Agassi's argument for the thesis that scientific theories are explanations of facts, I accept that thesis as a correct claim about the logical structure of scientific theories. The claim may be called the *explanationist* epistemology of science and may be supported on inductivist grounds.

chapter 9

Agassi's Philosophy of the Historiography of Science

AGASSI'S MAIN ARGUMENT in favor of his philosophy of the historiography of science, according to which philosophies of science are explanations of the facts of the development of science, is that it explains the following puzzling, interesting, and important facts in the development of the historiography of science: (1) "the study of the history of science is in a lamentable state" (p. v); (2) "historical explanation of any value is rare in the annals of the history of science" (p. viii); (3) "in the field of the history of science, divers efforts to study one and the same topic are not to be found anywhere, to my knowledge, with the exception of the studies of the history of Renaissance mechanics from those of Duhem and Mach to those of Koyré and his followers" (p. 75); (4) [there is a] "low standard of work on the history of science" (p. 77); (5) "a mediocre political history is

142

less boring than a mediocre history of science—even for a person like myself who is much more interested in the history of science" (p. 77); (6) "Bacon's problematic theory [that 'every scientific theory emerges from the facts' (p. 77) is] accepted as trivially true by so many historians of science" (p. 77); (7) "their histories [i.e., the histories of inductivist historians of science] as well as histories based on obviously inconsistent mixtures of inductivist and conventionalist views (see section 10) [are] so patiently tolerated by a rather enlightened public" (p. 77).

Agassi's explanation of these supposed facts constitutes the argument supporting his philosophy of the historiography of science. For the argument to succeed, the explanations must be acceptable in themselves and appropriately related to the philosophy. To determine the acceptability of the explanations and the appropriateness of the relation, a logical analysis of the explanations is necessary. This analysis[1] shall be conducted by using Agassi's own standards. The relevant standards are here what he calls "Popper's so-called deductive model of explanation, and . . . its application to history" (p. 75), which Agassi accepts, expounds, and uses in the last section of his essay in order to analyze and criticize history-of-science explanations.

Popper's deductive model,[2] which is somewhat similar to Hempel's, may be briefly stated: An explanation is a set of statements one of which, the *explicandum,* describes the state of affairs to be explained, while the other statements, the *explicans,* do the explaining. Assuming the explicandum to be true, the satisfactoriness of the explanation reduces to that of the explicans. The explicans may or may not logically entail the explicandum; if it does, then the explicans is satisfactory if and only if it contains universal statements and is true. If the explicans does not logically entail the explicandum, then the explicans may be either fully stated or merely sketched; if it is fully stated (since it does not entail the explicandum), it is unsatisfactory; if it is merely sketched, then its satisfactoriness depends on whether what has been omitted is trivially true or not; if it is not, then the explanation is not satisfactory, otherwise it is. The following are then necessary conditions for an explicans to be satisfactory: (1) it must be true, (2) when fully stated it must logically entail the explicandum, (3) when fully stated it must be testable independently of the explicandum by containing universal statements (otherwise the explicans is ad hoc or circular), and (4) when sketched, what is omitted must be trivial. The universal statements are often called "universal laws," though perhaps they should be so called only if they are true. The singular statements in the explicans are called "initial conditions."

Since some of my critical analysis depends on what Agassi explicitly says or does not say, I shall have to quote the explanations as he gives them. When Agassi explicitly describes a set of statements of his as an explanation, as he does in the case of explicanda (3) through (7) quoted above, there is little problem. But Agassi does not do so in cases (1) and (2). Hence I shall examine (3) through (7) first, the explanations being given in the last section of Agassi's essay.

Agassi's first explanation there is the following:

> In the field of the history of science, diverse efforts to study one and the same topic are not to be found anywhere, to my knowledge, with the exception of the studies of the history of Renaissance mechanics from those of Duhem and Mach to those of Koyré and his followers. I wish to explain this defect by the following hypothesis: although the laws used in historical explanation are usually very simple, where the history of science is concerned they are highly problematic and complicated. [P. 75]

The first step in a Popperian analysis is to separate what is being explained from what does the explaining. The explicandum is the first of the two statements in the set. The explicans is the "hypothesis" that:

(IC1) Although the laws used in historical explanation are usually very simple, where the history of science is concerned they are highly problematic and complicated.

This seems to be a singular statement of fact, about the logical properties of the concept of explanation. There is, however, no universal law in sight. The explanation is, then, not fully stated but merely sketched. Following Popper's theory, this is no criticism of that explanation. Agassi may or may not be justified in merely sketching his explanation, depending on whether or not the universal statement(s) he uses is (are) trivial. To evaluate Agassi's first explanation by his own standards we must determine what universal statement(s) he uses.

Leaving aside the problem of what is or could be the sense in which Agassi is "using" the universal statements, the question of *which* ones he is using is answered as follows. Agassi is using those universals which are contained in the "fully stated explanation" (p. 76). From Agassi's sketch one must construct the fully stated explanation. But what is such an entity? It is doubtful that there could really be fully stated explanations, and perhaps this notion does not even make sense. Of course, the notion can be given a sense, and this is what Popper's deductive model does. A fully stated explanation is one in accordance with the model, that is, one in which the explicandum deductively follows from a set of statements containing universal as well as singular statements (p. 75). We now know how to recognize fully stated explanations; the question remains, "How do we reconstruct them from the sketch?" And this construction is our present problem. This crucial question the model does not ask and, unfortunately, does not even seem to raise. In general there is no reason why there should be a one-to-one correspondence between sketches and fully stated explanations; there may very well be several ways of filling the sketch. This is admitted, implicitly, by Popper himself when he claims that "a clearer analysis" of the example of the breaking of a thread used by him "distinguishes *two* universal laws" instead of the one contained in his original analysis, though Popper does not seem to realize the consequences of that admission.[3]

Since our concern with the fully stated explanation originates from the problem of determining whether the universals contained therein are trivially true, the difficulty can perhaps be solved by taking a second look at the triviality question in the light of the difficulties encountered with the reconstruction problem. The triviality question is unambiguous only if the fully stated explanation to be reconstructed contains just one universal statement. In that case the question can only refer to whether the statement is uncontroversial or its plausibility trivially obvious. But, allowing the possibility that the fully stated explanation should contain more than one universal statement, then what may or may not also be trivial is the way in which these statements are to be combined with each other and with the initial conditions to yield the explicandum. For example, the truth of each universal statement may be trivially obvious, but the way in which they can be combined may be highly controversial or totally unknown. Therefore, one's Popperian analysis of an explanation sketch, such as Agassi's, is being more sympathetic if one "fully states it" with the help of one universal statement connecting the initial conditions with the explicandum. For in this case the only way in which the original sketch may turn out to be unsatisfactory is if the truth of the statement is not trivially obvious (and not, in addition, if the way of combining the many universals is not trivial). Moreover, it could not really be fair to expect the Popperian reconstructionist to "fill the sketch" with the help of many laws, for then the reconstruction problem would be as difficult as the explanatory problem itself. Unless the reconstruction problem is trivialized by requiring only one law, one could not use the model in the way that Popper does in *The Poverty of Historicism*, for example. Finally there is no possibility that the single law reconstruction be unfair, for the following reason. If this statement, which may be called the minimal universal, is trivially true, then the original explanation (sketch) will not be criticized for omitting it. But if that statement is not trivially true, then no set of laws contained in any multiple-laws reconstruction could be trivially obvious either, because that set entails the minimal universal and, if the latter is not trivial, either the content of at least one member of the set or their combination must be nontrivial. For all these reasons, it will only be fair, according to his own standards, to complete Agassi's explanations with only one connecting universal statement.

The universal statement being used by Agassi in his first explanation is perhaps the following:

> (L1) Whenever the principles used in a given field are highly problematic and complicated, then diverse efforts to study one and the same topic are rare in that field.

I said "perhaps" because the following generalization would also meet the required condition:

> (L1′) Whenever the laws used in explanations in a given field are highly problematic and complicated, then diverse efforts to study one and the same topic are rare in that field.

Finally there is a third possibility, though there is something rather artificial about it:

> (L1″) Whenever the laws used in history-of-science explanations are, unlike those used in ordinary historical explanation, highly problematic and complicated, then in the field of history of science diverse efforts to study one and the same topic are rare.

Since it is obvious that the explicandum follows logically from the conjunction of the initial condition (IC1) and either (L1), (L1′), or (L1″), an obvious difficulty arises, namely, that even a single-generalization reconstruction of an explanation sketch is not unique. And this seems to be true in general, since, if the explicandum of an explanation follows from the initial conditions and some universal statement, then it also follows from those initial conditions and any more general statement. If the deductive model is to be applicable, the question that has to be answered is, How general a universal statement is one supposed to construct when analyzing an explanation that has been merely sketched?

The criterion of independent testability is useless since a universal statement is automatically testable independently of the explicandum. This is true even for such an artificial generalization as (L1″), which is universal only with respect to time.

Let us now consider the criterion of maximum testability or falsifiability. This would allow us to use the most general universal statement which would fulfill the other requirements, in our case (L1). But this criterion is unacceptable because it is not evaluatively neutral. For the explanation being analyzed either is or is not a discovery. If it is a discovery, then to maximize the testability of the universal statement being used would be to attribute a desirable characteristic to the explanation. If the explanation is not a discovery, then whatever universal statement is being used, it is known to be true or false; in this case to attribute maximum falsifiability is to maximize the chance that the statement, and thus the explanation, is false. And to do this is to criticize the explanation.

It is not my task here to solve this difficulty, since what I intend to do is to criticize Agassi's explanations and the deductive model, each in terms of the other. At any rate, for my criticism of Agassi's first explanation, it is not necessary to solve the above mentioned problem because none of the three above mentioned universal statements is trivially true; and the same argument shows that all three of them are false. This can be seen as follows. Let us assume that the "laws" used in history-of-science explanation are indeed highly problematic and complicated. Consider now what will happen when a certain historical development is taken as an explicandum. Whenever some historian of science will put forth an explanation, the admitted problematic character and complexity of the universal statement or statements which he uses in his explanation will make the explanation itself problematic. These problems will be the source of dissatisfaction to other historians who will therefore be led to put forth a different explanation which in their judgment is more satisfactory.

They too will necessarily have been using problematic laws, so that the proliferation of explanations will go on indefinitely. And the proliferation is the direct result of the problematic character of the laws used in the study of one and the same topic. Our conclusion is that Agassi's first explanation is false.

Agassi's statement of his second explanation is as follows:

> Although this [i.e., Agassi's account of the logic of the concept of explanation] need not be taken too seriously, it suffices, I think, to explain quite a number of facts in the history of history.
>
> For my own part, I am trying to explain the low standard of work on the history of science. Unlike most historians, historians of science must, in the main, use laws that are highly problematical from methodology and epistemology. And unlike most historians who have to use problematical laws, they are seldom conscious of the fact that their laws are indeed problematical. [P. 77]

The explicandum is here "the low standard of work on the history of science." The explicans may be paraphrased without change of meaning as follows:

(IC2) Unlike most historians who have to use problematical laws, historians of science are seldom conscious of the fact that their laws are indeed problematical laws from methodology and epistemology.

This statement seems to be a generalization about what historians of science do, do not do, or have to do. If this is correct, then what would be missing from Agassi's second explanation sketch would be the initial conditions. But when we try to complete this sketch, we discover that what is needed is another generalization:

(L2) Whenever historians of science, unlike most historians who have to use problematical laws, are not conscious of the fact that their laws are indeed problematical laws from methodology and epistemology, then a low standard of work on the history of science results.

Which general statement is the "law" being used by Agassi in his second explanation? The answer is, of course, that the second generalization, (L2), is the one. Why? The Popperian answer might be that it is so because the second generalization, (L2), is a statement which allows us to deduce the explicandum from (IC2) and thus to explain the former. But (IC2) also allows us to deduce the explicandum from (L2). It seems, then, that the Popperian cannot really justify his choice of (L2) instead of (IC2) as the universal "law" being used in Agassi's second explanation sketch, for neither generality nor deductive power provides such a justification; the former does not because both statements are generalizations, the latter because both statements are necessary but insufficient for the deduction of the explicandum. I believe there is a reason why the Popperian would choose (L2) as the "universal law" being used by Agassi in his second explanation sketch: namely, (L2) states a connection (causal?) between the explicandum and (IC2), whereas (IC2) does not state a connection between the explicandum and (L2). But this is an un-Popperian justification because it shows that the function of what the adherent to the deductive model calls "universal laws" is to provide a connection between the explicandum and what does the explaining in the explanation sketch. But if that is the function of the universal law, then anything else that provides the connection is just as good. Universal laws, then, and perhaps even deduction would be accidents in explanation, and would not express its essence, let alone its meaning, as it is claimed by Popper.[4]

From the present point of view, then, (IC2) is to be regarded as a singular statement of sociological fact about historians of science and (L2) as a universal law. There still remains the question of whether (L2) is trivially true or not. Its nontriviality is shown by its falsehood. For the paradox of methodology is that, in general, one does not have to know explicitly what one is doing in order to do it, and even do it well. Agassi himself admits this when he says, "I do not think that *conscious* methodological efforts are *necessary* for a historian of science to write an interesting and valuable work" (p. 78, my italics). Thus Agassi's second explanation, besides being unjustifiably sketched, is also false.

Let us now examine Agassi's third explanation:

> Admittedly, bad or boring histories exist in every field, and no elaborate effort to explain them is necessary. Yet the fact is that a mediocre political history is less boring than a mediocre history of science—even for a person like myself who is much more interested in the history of science. Since this fact is rather puzzling, I do not think that my explanation is quite redundant. [P. 77]

The explicandum seems to be the statement that, even for a person much more interested in the history of science, a mediocre political history is less boring than a mediocre history of science (p. 77). The explicans is presumably the same as the explicans of the "low standard of work on the history of science" (p. 77). The universal statement that Agassi is using in his third explanation is perhaps the following:

> (L3) Whenever problematic laws must be used in a certain field, and the practitioners in that field are not aware of this fact, then the writings of these practitioners are boring, much more boring than the writings of investigators who either don't have to use problematic laws or are aware that the laws they use are problematic.

This statement is by no means trivially true; therefore according to the deductive model it should have been included in Agassi's third explanation, which is therefore unsatisfactory as it stands.

Yet the unjustified sketchiness of Agassi's third explanation may be the least of its drawbacks; it is probably false because the universal statement being used is probably false. This may be seen as follows. Since the laws used in political-history explanations are usually very simple, they will not normally be explicitly mentioned. Though the laws used in the explanations to be found in a history of science are problematical, since most historians of science are presumably not aware of this fact, they will not explicitly mention these laws either. Thus both political historians and science historians will take for granted and refrain from mentioning the laws they use. What is clear is that if both did explicitly mention the laws, the result would be that the political history would be much more boring than the history of science. The following is also clear; namely, if neither states the laws explicitly, the history of science is harder to comprehend, more provocative, and bolder. But if this is so, it would seem that the writings of the historian of science are more and not less interesting than those of the political historian. Thus (L3) is probably false.

Agassi's fourth explanation is the following:

> I would go further, and try to explain two more puzzles. First, why is Bacon's problematic theory ["according to which every scientific theory emerges from the facts" (p. 77)] accepted as trivially true by so many historians of science? . . . My answer . . . is this: Bacon's theory is the only well-known explanation of the rise of scientific ideas. Since the most obvious task of a

historian of the development of science is to explain one specific development, especially the development of an idea, he needs a general theory of the development of ideas; and there is no widely known competitor yet to Bacon's general theory of the emergence of scientific ideas. [Pp. 77–78]

The explicandum is the statement that "Bacon's problematical theory [is] accepted as trivially true by so many historians of science." The explicans is the following set of statements: "Bacon's theory is the only well-known explanation of the rise of scientific ideas. Since the most obvious task of a historian of the development of science is to explain one specific development, especially the development of an idea, he needs a general theory of the development of ideas; and there is no widely known competitor yet to Bacon's general theory of the emergence of scientific ideas." Does the explicandum follow from the explicans? To determine this, consider the following argument which exhibits the logical structure of the explanation more clearly.

(1) Bacon's theory is the only well-known explanation of the rise of scientific ideas.

(2) The most obvious task of a historian of the development of science is to explain one specific development, especially the development of an idea.

(3) If the historian of science wants to explain one specific development, especially the development of an idea, he needs a general theory of the development of ideas.

(4) Therefore, the historian of science needs a general theory of the development of ideas.

(5) But there is no widely known competitor yet to Bacon's general theory of the emergence of scientific ideas.

(6) Therefore, the historian of science will be likely to accept Bacon's general theory of the emergence of scientific ideas.

(7) Therefore, Bacon's theory will be accepted as trivially true by a great many historians of science.

Agassi's fourth explanation is then, at least in intention, "fully stated" and not "merely sketched"; for the presumption is that the last statement follows from the preceding ones, all of which are more or less explicitly stated by Agassi himself, and among which one finds a universal statement, namely (3). It is doubtful, however, whether this intentionally full statement makes the explanation any more satisfactory, for it now becomes open to other kinds of objections.

First, even if (6) should follow from the preceding statements, (7) does not follow from (6). Nor is the difference between these two statements trivial. In fact, according to Agassi, the difference between "acceptance" and "acceptance as trivially true" is the difference between "acceptance" and "uncritical acceptance," and it is primarily the latter that needs to be criticized and explained. It is not difficult to understand how Agassi was led to such an oversight. We recall that Agassi is using this explanation to boost the explanatory power of his philosophy of the historiography of science, of which an important part is the deductive model of explanation, a presumed consequence of which model is the universal statement (3) in the argument. The direction of his thinking is then from theory to explicandum and not from explicandum to explicans. Looked at from the point of view of the theory or statement (3), the differences between the last two statements (6) and (7) do not appear to be great: they may both be regarded as confirming the former.

The second objection is this: (6) is being fallaciously deduced by Agassi from (4) and (5). The subterfuge is the identification of the "general theory of the development of ideas" in (4) with the "general theory of the emergence of scientific ideas" in (5). By comparing the statements from which (4) and (5) are respectively deduced, we can see how different those two things really are. (4) is validly being deduced from (2) and (3); hence the "general theory of the development of ideas" mentioned in (4) must be the one mentioned in (3); and the "general theory of the development of ideas" in (3) is the theory of the development of ideas in general, needed if one wants to explain a specific development of an idea; in other words, it is the "universal law" (about how ideas develop) being used, in accordance with the deductive model, in the explanation. But the "general theory of the emergence of scientific ideas" in (5) must be simply the theory about how scientific ideas originated in the first place, i.e., the account of the historical event called the rise of scientific ideas; and the reason for this is that Agassi deduces, fallaciously, as we shall see later, (5) from (1), and (1) speaks of the explanation of the historical event labeled "the rise of scientific ideas." Thus, since what the historian of science needs, according to the deductive model, is a theory about the development of ideas in general, and since what lacks widely known competitors is Bacon's theory of how scientific ideas originated, it does not follow that the historian will accept Bacon's theory.

I said above that Agassi is not justified in deducing (5) from (1). And this is my third objection. I think that Agassi must be trying to deduce (5) from (1), otherwise (1), which is adduced in his statement of the explanation as the principal initial condition, would be superfluous. Moreover, both (1) and (5) speak of what is clearly the same thing, namely, the "rise of scientific ideas" and the "emergence of scientific ideas," respectively. Agassi equates an explanation with a theory. And this is surely unjustified, even from the point of view of the deductive model which emphasizes the need for initial conditions. I believe that this misapplication of the deductive model stems from a misapplication of "Popper's critical philosophy of science"; that is, Popper's idea that scientific theories are explanations of facts is effectively being extended by claiming that the converse is also the case, namely, that explanations of facts are scientific theories, i.e., explanations are general theories. And this extension is not justified since scientific theories are explanations of scientific facts, which are already of such a generality as to obviate the need for initial conditions in the explanation of them. But not all facts are scientific facts; the most obvious contrast is historical facts, and for the explanation of these, theories alone are not sufficient, and may even be unnecessary.

The above critical analysis has shown that Agassi's fourth explanation is highly unsatisfactory when judged by the standards of the deductive model of explanation. To be sure, since the model applies only to fully stated explanations, and since Agassi's explanation did not come with the label "fully stated," that evaluation of unsatisfactoriness depends also on our decision, not unfounded, to regard this explanation of Agassi's as fully stated. Perhaps that decision was mistaken. Let us then take the set of statements constituting the explicans not as an intentionally full statement of the grounds from which the explanandum is supposed to follow, but rather as a sketch which perhaps remains to be filled. The statements following the main initial condition of the explicans will be regarded, then, as an articulation of the meaning of that initial condition. And Agassi's equation of "explanation" with "theory" and of "rise of scientific ideas" with "development of scientific ideas" will be regarded not as fallacious deductions, but as Agassian nominal definitions. When all this is done, Agassi's fourth explanation becomes a sketch in which the explicans is the statement that Bacon's theory is the only theory of the develop-

153

ment of scientific ideas well-known among historians of science, and this is supposed to explain why Bacon's theory is accepted as trivially true by so many historians of science. But, if this is so, and if we do not emphasize the *uncritical acceptance* of which the explicandum speaks, then the explanation is circular since both explicans and explicandum mean the same thing; the explanandum does indeed follow from an explicans which does not contain a universal statement. Thus, in any case, Agassi's fourth explanation is unsatisfactory, according to the deductive model.

The fifth explicandum, the alleged fact that certain histories are tolerated by an enlightened public, is not really a historical fact about historiography of science, but about one of the developments of Western culture, since it consists of the public's attitude toward the historiography of science. Hence its explanation is not likely to be relevant to Agassi's philosophy of historiography of science. I shall not, therefore, examine it.

This part of Agassi's argument is then without force, since the explanations constituting it are unsatisfactory. This fact makes it unnecessary to examine in detail the question of what elements of that philosophy are so related to parts of those explanations as to have their explanatory power increased. A secondary result of our analysis is the exhibition of the difficulty of applying the deductive model and the problematic character of many of its aspects.

I now turn to the other part of the argument, where Agassi deplores the lamentable state of the history of science, which I quoted in chapter 8. Two interconnected questions arise. First, is there really anything to explain, and second, does Agassi really put forth an explanation? Both questions are disposed of by the deductive model of explanation (a good reason for not accepting the model, I should say). Let us answer the second question first. It is true that Agassi is not explicitly calling this set of statements an explanation, as he does his statements in the last section of his essay, which we have already examined. But, according to the deductive model, an explanation is a set of statements one of which follows logically from the others, which ideally consist of singular statements of initial conditions and of universal laws. Thus any argument which attempts to prove a conclusion is also an explanation. Therefore, since in the above quotation Agassi is obviously sketching an argument intended to show that the study of the history of science is in a lamentable state, he is also sketching an explanation of that alleged fact. Moreover, he is obviously using causal language in "the faults which have given rise to this situation . . . stem . . . ," and causal claims are, according to the deductive model, sketched (causal) explanations.

This answer to the second question helps us to answer the first, which asks whether it is appropriate to look for an explanation of that alleged fact, when we do not even know whether it is really a fact. The answer is that it is perfectly appropriate to do so since the explanation will be the proof of the fact.

But there is another problem that must be faced now, in that the explanation being considered is the intended proof of a dubious fact. I also interpret the present explanation to be part of Agassi's argument in support of his philosophy of the historiography of science. I am then interpreting this explanation as an argument which supports both the historical fact which is presumably its conclusion, and the philosophical claims which are presumably its premises. Am I not thereby attributing to Agassi's argument a vicious circularity, which perhaps reflects more on my interpretation than on his statements? The answer to this question is two-fold. First, the circularity, to the extent that there is one, must be attributed to Agassi himself, since I have already shown that the present interpretation is the most proper and the most consistent with his general outlook. Second, the circularity involved may not be as objectionable as it may at first appear, if the circumference of the conceptual circle involved is long enough to allow one to gain a great deal of understanding in the process of going around.

Consider the structure of Agassi's explanation-argument for the lamentable state of history of science:

(1) Most or many historians of science accept uncritically the inductivist or conventionalist philosophies of science.

(2) The acceptance of these philosophies leads the historian to write inductivist and conventionalist histories, respectively. This is so because philosophies of science are explanations of the facts of the history of science, so that the latter are derived from the former, the acceptance of which thus becomes the intellectual cause of those histories.

(3) There is nothing objectionable if the historian of science accepts those philosophies of science "critically"; for then the study of the history of science would be the endeavor to explain facts by means of those philosophies, and Agassi's phi-

losophy of the historiography of science assures us that philosophies of science are explanations of the facts of the development of science.

(4) But for the historian of science to accept those philosophies
uncritically is irrational because the acceptance or allegation
of facts without any connection to a theory is at best, if the
theory is true, an activity without purpose or aim. At worst,
if the relevant theory is false, it is useless for the advance-
ment of learning, which would require first a correction or
articulation of the theory and then new kinds of facts to cor-
roborate it, instead of more confirmatory facts of the old kind,
which do not really add much to the rational acceptance of
the theory. Again, it is Agassi's philosophy of the historiography
of science which assures us that the relevant theories are philosophies of science.

(5) Therefore the study of the history of science is in a lamentable
state.

Although premise (1) is a claim about present historians, it is a
claim about the philosophies of science they accept and not specifically
about the kind of history they write. Reference to the kind of history
they write is made in Agassi's conclusion. Agassi supports premise
(1) by verbal evidence such as McKie's dogmatic statement that scientific ideas emerge from facts (p. 6). This is proper evidence for
the kind of claim Agassi is here making, which is a claim about the
philosophy of science accepted by historians.

Premise (2) is put forth primarily as a consequence of Agassi's
philosophy of the historiography of science, and it is presumably
corroborated by Agassi's interpretation of the historical developments
adduced by him in his essay. I think, however, that Agassi misinterprets his own evidence here. At best, this evidence shows that during the golden age of inductivism historians of science were led to
write inductivist histories by their acceptance of the inductivist philosophy of science, and that Duhem was led to write conventionalist
history by his acceptance of the conventionalist philosophy of science.
But this does not mean that historians are now led to write inductivist
and conventionalist histories because of their respective acceptance
of those philosophies. At any rate what Agassi's evidence really shows
is that it was the historian's extension of the inductivist and conventionalist philosophies of science into the domain of history that

led him to write inductivist and conventionalist histories, and not the simple acceptance of a given philosophy. And it may be noted that in this respect Agassi is doing exactly what inductivist and conventionalist historians did, namely, extend the Popperian philosophy of science into the field of history of science.

Claims (3) and (4) are derived essentially from Agassi's philosophy of the historiography of science; we have seen that this is also the case with premise (2). Thus Agassi's philosophy does have some explanatory power. But this explanatory power is insignificant; the above mentioned conceptual circle is too small. For, if the only "lamentability" in history of science is the alleged "irrationality" mentioned in (4), then Agassi's present explanation is ad hoc.

Even more ad hoc is the explanation constituting Agassi's brief summary of the essay's final section in which are contained the first five specific explanations already examined: *"Historical explanation of any value is rare in the annals of the history of science, mainly because of a naive acceptance of untenable philosophical principles"* (p. viii). What is Agassi's evidence that "historical explanation of any value is rare in the annals of the history of science?" It is true that he has given much evidence intended to show that there is among some historians of science a tendency to transcribe more details than they can either check or understand properly, to be "wise after the event," to use indiscriminately and irresponsibly the "continuity theory" and the "emergence technique," etc. But this at best shows that other aspects of historical practice are unsatisfactory; it has little to do specifically with *explanation*.

Agassi does have one Popperian argument: All fully stated explanations must contain universal laws, otherwise they are ad hoc or circular; and explanations must be fully stated when the universal laws are nontrivial. The universal laws of the historian of science are nontrivial; therefore those explanations should be fully stated. Since they are usually not fully stated they are unsatisfactory.

But what is Agassi's evidence that historians of science naively accept the questionable principles used in their explanations? It is the fact that they do not mention those principles in their explanations, i.e., that their explanations are unsatisfactory in the above sense. Agassi's own explanation is then worse than ad hoc, it is circular. The conclusion of this examination of Agassi's main argument is that he has provided no good reasons for accepting his philosophy of the historiography of science, while some reasons for abandoning it have been found.

chapter 10

History-of-Science Explanation:
Its Autonomy

Inductivist Argument for Agassi's Idea

ONE OF AGASSI'S conjectures examined in the last chapter, if interpreted as a solution to the problem of explanation in the historiography of science, has, on an inductivist interpretation, much initial plausibility and, at least for philosophers of science, great appeal: "historical explanation of any value is rare in the annals of the history of science, mainly because of a naive acceptance of untenable philosophical principles," [1] that is to say, because "unlike most historians, historians of science must in the main use laws that are highly problematical from methodology and epistemology [and] unlike most historians who have to use problematical laws they are seldom conscious of the fact that their laws are indeed problematical." [2] As an example of such a philosophical principle, and one which is dogmatically accepted and presumably employed in explanation by some inductivists Agassi cites "Bacon's law according to which every scientific theory emerges from the facts." [3] The practical consequences of Agassi's conjecture when so interpreted are clear: since the cause of the unsatisfactory state of history-of-science explanation is that historians of science uncritically accept principles of the methodology and epistemology of science, they should, if they want to improve their explanatory practice, become more critical about the philosophy of science principles they accept. And they can do this by taking seriously, studying, and critically evaluating the works of philosophers of science in Agassi's sense, that is to say, philosophers of the history of science.

However, if Agassi wants to accept his own thesis critically, if he wants to have reasons for (or to justify) his belief, he must do it as an inductivist. He must make his idea "emerge" from the facts of contemporary historiography of science. For we have seen that he cannot accept that thesis on his own Popperian or hypothetico-deductivist grounds since it is ad hoc and circular. And yet the hypothesis clearly is a recurring theme throughout Agassi's essay and one of its principal messages. On the other hand, once an inductivist approach is adopted in the field of the philosophy of the historiography of science—that is, for Agassi, in the domain of facts of the development of the historiography of science—the possibility immediately arises that inductivism may be the best approach, not in science, to be sure, but in history, including history of science, i.e., in the domain of facts of the development of science, which the Agassian philosopher of science studies no less than does the historian of science. Agassi's neglect of this possibility must be due to his having succumbed to the common tendency of trying to imitate science by transferring its methods to history, a tendency which, ironically enough, he derisively attributes to inductivist and conventionalist historians of science.

But how can Agassi's idea be made to emerge from the facts of the historiography of science, which I established in Part I? The argument would run as follows. We know for a fact that history-of-science explanation is in an unsatisfactory state, for which there must be a good reason. A possible cause is the historians' uncritical acceptance of philosophy of science principles. Nothing else seems to be present which could account for this deficiency. Therefore its explanation lies in their uncritical acceptance of philosophical principles. That the argument is sound can be seen by exhibiting its structure:

(1) *E* is unsatisfactory.
(2) Everything must have a cause.
(3) *C* is a possible cause of *E*.
(4) There is no other cause of *E* present.
(5) Therefore, *C* is the cause of *E*.

It is true, of course, that in order to derive (5) we need (2), (3), and (4) as well as (1). But it is clear that, in arguments of this form, (1) is the most important claim in the sense that it is the claim which gets that process of reasoning started; moreover (2) is normally presupposed by everyone; (4) is often established simply by inspection; and (3) is often intuitively obvious. Since I have established (1), we shall have to accept Agassi's conjecture if it is the case that (3) must be accepted and we do not provide an alternative.

One comment that should be made about Agassi's conjecture is that it has the merit of stating the right type of explanation of the unsatisfactory state of history-of-science explanations. It does in fact state the presumed intellectual cause of the undesirable state of affairs, and to do so is in the best historical and philosophical tradition. But do the merits of Agassi's conjecture extend beyond this formal propriety? Since the proposition corresponding to (3) is not intuitively obvious in our case, what we must determine now is whether the historians' uncritical acceptance of principles of the philosophy of science can indeed account for the deficiency of history-of-science explanation.

Application to Guerlac's Explanation

Let us return to Guerlac's explanation of Lavoisier's autumn 1772 experiments and consider whether it fails because Guerlac is unaware that his underlying philosophy of science principles are problematical and controversial.

In *Lavoisier—The Crucial Year* Guerlac uses all sorts of problematic principles, even philosophical ones. But the ones to which Agassi refers are philosophy of science principles. Three examples Agassi mentions are "the *inductive philosophy* of science, according to which scientific theories emerge from facts, . . . the *conventionalist philosophy* of science, according to which scientific theories are mathematical pigeon-holes for classifying facts . . . [and] Popper's *critical philosophy* of science . . . [on which] view scientific theories explain known facts and are refutable by new facts." [4] If we think in these terms, Guerlac may seem to be using the inductive philosophy. For he takes great pains to show (without success, to be sure) that the weight-augmentation effect had been established as a fact and so accepted by Lavoisier previous to his conception of the antiphlogistic hypothesis; and Guerlac also argues (again unsuccessfully) that two other facts, limited calcination in closed vessels and the effervescence of calxes during reduction, contributed to Lavoisier's development of the hypothesis. However, though Guerlac does emphasize these facts, he attaches no less importance to Lavoisier's *idée maitresse,* that air plays a chemical role. His belief that Lavoisier's antiphlogistic hypothesis originated from that idea forces him to assume Lavoisier's knowledge of the above mentioned facts, which he sees as combining with that idea to produce Lavoisier's hypothesis. Moreover, in a superficial sense, Guerlac is also using Popper's critical philosophy. For he characterizes Lavoisier's antiphlogistic hypothesis primarily as an explanation of the augmentation effect. Thus if Guerlac is using any principle in his explanation, it is the principle that scientific ideas emerge from the appropriate facts and ideas and they are explanations of these facts from which they emerge. But the very wording of this "principle" makes uncritical acceptance of it impossible, since the judgment of the appropriateness of the facts and ideas and of whether the new ideas explain the facts presupposes critical evaluation. In fact, it would be misleading to call the above mentioned statement a principle at all. It follows that Guerlac could not have been led to his error by his unawareness of the problematic character of any of the three principles of the philosophy of science mentioned by Agassi.

The above argument is however superficial and criticizes only Agassi's letter, not his spirit. If we want to criticize Agassi's basic approach, we must recognize that, on an inductivist view, the facts from which scientific ideas emerge are things termed *facts* by the latest textbook. Thus the "idea" that air plays a chemical role is really a fact. In other words, according to inductivism, scientific knowledge is the body of knowledge constituting the latest science textbook, and the history of science is the temporal evolution of the content of that textbook. The inductivist principle of scientific development is then the principle that scientific ideas or facts emerge or originate from previous scientific ideas or facts, where *scientific* means "found in the latest text," that is, "factually or substantively correct (even if approximately so)."

When we understand inductivism in this manner, it cannot be denied that Guerlac could have been led to his error by his acceptance of the inductivist philosophy. For when Guerlac sets out to explain, or find the origin of, Lavoisier's experiments referred to in the famous sealed note, he conceives his task to be that of finding evidence that, previous to those experiments, Lavoisier had the appropriate scientific (i.e., "correct") beliefs and knowledge. Which beliefs and knowledge are appropriate? They are the ones to which experiments of the kind made by Lavoisier can be connected scientifically, that is, correctly by the standards of contemporary scientific knowledge, i.e., on the basis of the textbook knowledge. Or alternatively, to describe the temporal sequence of Guerlac's own researches,[5] when he thinks he has found evidence that previous to the sealed-note experiments Lavoisier had at least some of the appropriate scientific beliefs and knowledge, he thinks he has found the origin of those experiments and has the basis for a causal explanation of them.

That is why Guerlac, on reading Lavoisier's August memorandum, ignored the latter's thinking in terms of the phlogiston and why he misinterprets Lavoisier's belief that air is released by metals when heated. To do otherwise would be to contaminate Lavoisier's scientific thinking with incorrect and thus unscientific beliefs.

The above argument does not show that Agassi's conjecture is correct in Guerlac's case, however. For it is not Guerlac's uncritical acceptance of the inductive principle that led him to error. It seems that as long as Guerlac accepted the principle, whether critically or uncritically, he could not have avoided the error. Or to be more exact, he could have avoided it only by avoiding the investigation of Lavoisier's discovery altogether, by not applying the principle in this case. Can we then say that the critical acceptance of the principle would have led Guerlac to avoid the investigation of Lavoisier's experiments? It is doubtful that even this can be said; for how could Guerlac know in advance that the present case is one where his principle cannot be applied and thus one not to be investigated? It seems that Guerlac would have had first to "establish the facts" of the case, or to put it differently, to first find the explanation of Lavoisier's experiments.

It would be irrelevant to argue, as Agassi might be inclined to, that I am here assuming gratuitously that there are facts to be established and that they can be established by so wanting without using any principles. In fact, all I need for my argument is the existence of purported facts and the possibility of establishing such purported facts without principles. For I am arguing that the critical acceptance of the inductive principle would have led Guerlac either to the same error or to avoid the application of that principle on the grounds that he believed that the facts of the present case made such an application impossible.

Moreover, if Agassi does not regard it as an established fact that the inductive principle is not applicable to Lavoisier's discovery of the explanation of the augmentation effect, he cannot logically accept the fact that Guerlac's explanation is unsatisfactory; and if he does not accept it, he thereby gives up the hope of making his own conjecture emerge from the facts of the historiography of science, namely, of providing an argument in favor of his conjecture.

Finally, if we take seriously Agassi's notion of the critical acceptance of a philosophy as the explicit acceptance of a philosophy with the full awareness of its problematic character, then it is likely that it would be a critical acceptance of inductivism that could lead to an error such as Guerlac's. For the self-conscious application of the inductive principle to recalcitrant cases like Lavoisier's discovery would lead to a distortion of the facts and a misinterpretation of the evidence, in order to show that the principle is not, after all, refuted by the problematic case. Our conclusion must be that it is the mere acceptance, and not the uncritical acceptance, of the inductivist philosophy that could have led Guerlac to failure.

Application to Koyré's Explanation

It is also true that Koyré could have been led to his unsatisfactory explanation by his acceptance of a certain philosophy of science. This philosophy may be called, for lack of a better term, *Archimedean* since it can be briefly stated by saying that modern (classical) science is Archimedean.[6] Modern science is then, according to Koyré, Archimedean, not in content, obviously, but in method. In fact, according to this philosophy of science, it is too naive, or at least not sufficient, to characterize science in terms of the body of doctrines contained in the latest textbook; science must be characterized in terms of its method, that is, its approach to the study of nature. This scientific method is nothing but the quantitative, as contrasted to the qualitative approach to the investigation of natural phenomena; for the world of modern classical science is indeed a world of quantity and not of quality. Now, at this point the Archimedean philosopher of science becomes curiously inconsistent. For him Archimedeanism is supposedly one of the two historically significant strands of Platonism (the other being "mystical arithmology" [7]), and is to be contrasted to Aristotelianism. But here he adopts, curiously enough, the Aristotelian philosophy of mathematics, according to which mathematics is the study of quantity and number. Thus the scientific method is further characterized as the mathematical approach to nature. The history of modern science thus becomes the temporal evolution of the attempts, unsuccessful as well as successful, to describe and explain everything quantitatively and mathematically.

The rise of modern science is the beginning of the unfolding of the substantive results from the use of the new method of mathematization. The founders of modern science thus had to criticize previous doctrines not only for their content but also for their method, and they had to defend the new approach as well as the new theories. And because they had to build the very framework which made their discoveries possible, we would expect that the discovery of laws like the law of falling bodies, which to us appears very simple, should be preceded by many failures. The discovery of the law of falling bodies in particular acquires special importance for the Archimedean philosopher of science. In fact, when formulated as the so-called law of odd numbers, the law "subjects motion to numbers." That formulation states that the distances traversed by a falling body in successive equal intervals of time are proportional to the odd numbers. If the Archimedean philosopher of science could show that the Galilean success in the discovery of the law was the result of Galileo's mathematical approach to nature, the Archimedean would have thus provided historical confirmation for the Archimedean philosophy of science, if not actual proof as Galileo himself and some of his disciples allegedly believed. In fact, Galileo at times seems to believe that his success with the law of falling bodies represents at once an experimental proof of the Archimedean philosophy of science and an experimental refutation of the Aristotelian qualitative philosophy of nature, a refutation in an area where the qualitative approach had seemed most secure and apt, namely in the phenomenon of motion, one which almost by definition seemed to defy quantification and one which even the divine Archimedes had been unable to mathematize.

The present-day philosophical consequences of the above mentioned and of other historical confirmations of the Archimedean philosophy of science cannot be overestimated. In fact, human behavior is the phenomenon that provides the present-day analog to Galileo's motion. Certainly the logical, methodological, and practical arguments implying the necessary inexactness of the study of human behavior are very impressive, and I am very much inclined to accept them. Yet who will say that their cogency exceeds that of the Aristotelian arguments against the possibility of an exact science of motion? I have shown, however, that Koyré's historical confirmation of the Archimedean philosophy of science is illusory.

Be that as it may, the Archimedean principle of scientific development is that scientific successes are the result of the correct use of the method of mathematizing nature, scientific failures are due to the incorrect use of the method.

Thus if Koyré wants to explain Descartes's failure and Galileo's early error he must find (or create) evidence that they misapplied the method, that, at a time when the only mathematization could be geometrization, they geometrized excessively, or that Descartes's mathematicism neglected the physical, or that when, after 1630, Descartes did pay attention to nature and the physical, he did not mathematize. And to explain Galileo's success Koyré must say and argue that it was due to his physico-mathematicism, his ability to mathematize nature.

Koyré's acceptance of the Archimedean philosophy of science, then, could have led him to his failure in the explanation of Galileo's success and Descartes's error. But does this show that Agassi's conjecture, or to be more exact, an Agassian approach is right? Agassi does not mention the Archimedean philosophy of science as a possible source of the deficiency of history-of-science explanation. Furthermore, the three principles that he does mention are of a different category. They may be called principles of the epistemology of science because they state the logical relations among various elements of scientific knowledge, specifically between scientific theories and facts. The inductive philosopher may be taken to assert that scientific theories are true (even if often only approximately so) generalizations derived from facts. The conventionalist philosopher asserts that scientific theories are not the sorts of things that can be true or false. They do not perform the same function as that of factual statements; for example, they are unfalsifiable because, given any prima facie refuting fact, one can always rearrange the theoretical framework or reinterpret the fact so as to eliminate the initial logical conflict between theory and fact. And if they are unfalsifiable it makes little sense to say that they are true; their function is in fact to provide useful classifications and organization of known facts and the prediction of new ones. The Popperian philosophy asserts that scientific theories are indeed not provable in any important sense, but that they are the sorts of things that could be true because they are falsifiable. They are, in fact, hypotheses from which facts may be deduced, thereby explaining the facts but not proving the hypotheses; alternatively, what is deduced may contradict a fact, thereby refuting the theory.

It is not clear whether Agassi means these principles to state the logical relations valid for the historical agents or those valid in reality, that is to say in the best judgment of the historian himself. This confusion is one that leads Agassi into difficulties, and many historians of science along with him. In spite of it, these principles, principles of the epistemology of science, can be distinguished from the Archimedean principle, which may be called a principle of the methodology of science. The latter are those that would state the causal relations among scientific successes or failures or errors and methods. For example, the Archimedean principle asserts that scientific theories are causally obtained by, are the result of, the successful use of the mathematical approach to nature. And successful uses are those that result in the theories constituting the latest textbook.

There are, of course, logical and empirical connections between epistemology and methodology of science in the above sense. For example, the inductive principle could be interpreted methodologically: not only are scientific theories generalizations derived from the facts, but they were possible because they were derived from the facts generalized upon. Moreover, it seems that most Archimedeans, including Koyré and Gillispie,[8] are inductivists, at least in the sense that for them scientific theories are those to be found in the latest textbooks.

But these details should not distract us from our present problem. Though Agassi does not mention the Archimedean principle, the attempt to explain Koyré's failure by reference to it is still Agassian in the sense that it blames the deficiency of historical explanatory practice on some principle of the philosophy, specifically the methodology, of science.

But we have also to determine whether Koyré's uncritical acceptance of that principle could have led him to his failure. Here we can argue, as we did in the case of Guerlac, that the uncritical acceptance could not be the cause since the critical acceptance would have produced, with even greater likelihood, the same effect. Moreover, it cannot be claimed that philosophical naiveté characterizes Koyré's attitude or practice. For, though in the second volume of the *Etudes Galiléennes,* which is the relevant work since it contains the explanation criticized by me, Koyré does not explicitly discuss and criticize other philosophies of science, as he does in his later essays of 1943 [9] and of 1948,[10] nevertheless the footnotes in that work show that Koyré is indeed aware of the fact that other philosophies of science could be used to interpret differently the historical materials he is using.

These differences of awareness of the philosophy of science that Koyré accepts at different times give the clue to what I think is the correct causal order. It was not his uncritical acceptance of the Archimedean principle that led him to put forth his untenable explanation, rather it was his mistaken belief in his explanation that led him to a critical acceptance of the Archimedean philosophy. And if it can be shown that such is the causal order in general, then we would have proved the dependence of Agassian philosophies (of the history) of science on history of science. But before chaining the theoretical history of science to the historiography of science, let me liberate the latter.

History-of-Science Explanation Liberated

I have argued that, insofar as philosophy of science principles are relevant to the deficiency of history-of-science explanation, it is the sheer acceptance of them that could be the cause of that unsatisfactoriness. If so, without determining whether the possibility is a reality, we could conclude that what the historian of science should be watchful for, what he should refrain from, is the acceptance of any philosophy of science "principles." At this point Agassi's claim enters: that historians of science need to use philosophy of science principles in and for their explanations. If this need is real, then the conclusion to be drawn from the above mentioned causal possibility is not that philosophy of science principles should be kept at a distance by historians of science, but that they should be used and accepted with the full awareness of their "problematic and controversial character."

The question to be settled is then whether or how philosophy of science principles are used or needed in history-of-science explanations. One possible use of these principles, which may be called "methodological," we have already mentioned; it is to find or construct explanations. In other words, those principles could be used to find evidence which explains certain aspects of the events of the development of science. To

167

be more specific and to repeat in part: given the problem of explaining Lavoisier's performance of his epoch-making experiments, the inductive principle tells us what kind of evidence we have to find or create. Similarly, if we want to explain the failure and successes of Descartes and Galileo in the discovery of the law of falling bodies, the Archimedean principle tells us to look for a mathematical aprioristic feature in their procedures.

It is clear, however, that no one of these principles need be used. This is so because, in a sense, all of them are needed simultaneously; and since, as general principles, they conflict with each other within each group (epistemology of science and methodology of science, respectively), they have to be reformulated so that they are no longer principles and do not conflict with each other. For example, the inductive principle should be reformulated as the inductive possibility that scientific theories may emerge from "facts" (what the latest textbook regards as facts); the Popperian principle as the hypothetico-deductive possibility that scientific theories may be conjectures explanatory of known facts and refutable by new ones; the Archimedean principle as the possibility that scientific successes may be due to the correct use of the mathematical approach to nature. All the historian needs is an awareness of the various possibilities; and this awareness need not be theoretical, it is sufficient that it be practical. That is, all the historian needs is to be able to recognize the case under investigation for what it is when he finds it. My criticism of Guerlac and Koyré was intended to show that they did not succeed in so doing in the case they investigated.

If all that the historian needs in this context is practical awareness of the various possibilities, could Agassi now argue that he is still methodologically dependent on the philosophy of science on the grounds that the relatively more speculative nature of the philosopher of science makes him better equipped to think of possibilities than the historian? I do not think so because the possibilities in question are not purely logical possibilities, but historical possibilities. For example, one speculative possibility, popular among certain philosophers of science, is that a scientific theory is an infinite set of sentences consisting of the deductive closure of a number of axioms, that is to say, a theory in the sense of mathematical logic. Such a philosophy of science principle seems to be of the same type as the inductive, conventionalist, or Popperian principles. Yet the methodological usefulness, let alone the necessity, for history-of-science explanation of such a logicist principle of the philosophy of science is rather doubtful to say the least. Moreover, it is simply a historical fact that the study of history often reveals possibilities undreamed of by the logician-speculator. Therefore, the historian of science in his explanations is not methodologically dependent on the philosophy of science, at least not on the philosophy of science as understood by Agassi.

The other possible use of philosophy of science principles by the historian of science is one for the purpose of justifying the explanation, once it has been found; this may be called the logical use. Agassi's argument for the logical need of philosophy of science principles in history-of-science explanation is a special case of a general argument in support of what I in chapter 3 called the Law-coverability Principle for Explanations.

Let us apply the argument to Guerlac's explanation. Guerlac explains Lavoisier's performance of the litharge experiment mentioned in the sealed note by saying that Lavoisier wanted to test his antiphlogistic hypothesis which asserts that the air absorbed by metals during calcination and released upon reduction can account for the greater weight of a calx. In turn, Guerlac explains Lavoisier's formulation of this hypothesis as follows: Lavoisier reached the antiphlogistic hypothesis because (1) he was convinced that air plays a chemical role in the phenomenon of calcination, (2) he knew that effervescence-like phenomena can be observed in the reduction of metallic calxes, (3) he was aware that metals resist calcination in closed vessels, and (4) he knew that metals which can be burned gain weight when they are calcined. Let us fix our attention on this last 'because' thesis; it is one, no doubt the principal one, of the many 'because' statements constituting Guerlac's explanation. If we had started with the problem of determining how Lavoisier came to formulate his antiphlogistic hypothesis, that 'because' statement would be its explanation. It thus reflects many of the analytical features possessed by Guerlac's explanation as a whole. In particular it provides an excellent illustration of Agassi's argument concerning the logical need of philosophy of science principles in history-of-science explanations.

Let us assume, for the sake of the argument, what I have in fact shown to be untenable, namely, that (1) through (4) are true. It is clear how one could justify or criticize the accuracy of those assertions. But how can Guerlac justify his claim that those assertions explain why Lavoisier formulated his antiphlogistic hypothesis? That is, how can Guerlac justify his claim, which is obviously essential to his ex-

169

planation, that the reasons mentioned in (1)–(4) were indeed effective, that they were in fact the motivating reasons? Lavoisier's conception of the antiphlogistic explanation may indeed have come temporally after Lavoisier's knowledge and beliefs (1)–(4). But how can one show that it came because of those beliefs and knowledge?

According to Agassi the only way of justifying the causal aspect of the explanatory thesis is to appeal to the following generalizations:

(C1) Whoever is convinced that air plays a chemical role in calcination and that metals resist calcination in closed vessels will conclude that metals resist calcination in closed vessels because of the relative lack of air.
(C2) Whoever knows that metallic calxes effervesce during reduction will conclude that metallic calxes give off air when reduced.
(C3) Whoever knows that metallic calxes give off air when reduced to metals and that metals resist calcination in closed vessels because of the relative lack of air will conclude that metals absorb air when burned or roasted.
(C4) Whoever knows that metals absorb air when burned or roasted and that all metals that can be calcined gain weight when changed into a calx will conclude that the gain in weight is due to the absorption of air during calcination.

These four generalizations in turn imply the following:

(C) Whoever is convinced that air plays a chemical role in calcination, that metals resist calcination in closed vessels, that metallic calxes effervesce during reduction, and that all metals that can be calcined gain weight when changed into the calx will conclude that the gain in weight is due to the absorption of air during calcination.

According to Agassi, it is only this generalization (C) that can justify our saying that Lavoisier's antiphlogistic conception not only followed his four previous beliefs but came into being because of those four beliefs of his. But, so Agassi might continue, the generalization (C) is a principle of the philosophy of science. It is in fact a special case of the inductive principle. Moreover, (C) is only a psychologistic way of saying that the following argument is logically valid:

(A1) Air plays a chemical role in calcination.

(A2) Metals resist calcination in closed vessels.

(A) (A3) Metallic calxes effervesce during reduction.

(A4) All metals that can be calcined gain weight when changed into the calx.

(A5) Therefore, the gain in weight by metals during calcination is due to the air absorbed in the process.

And questions concerning the logical validity of arguments are the special province of the philosopher.

Agassi's argument may be criticized as follows. It is at best misleading to call (C) a philosophy of science principle since it suggests that the philosopher of science is more competent to deal with it than the historian of science. But though it is true that logical questions tend to concern the philosopher more than the historian, the logical validity of (A) in the abstract with the consequent truth of (C) in general is utterly irrelevant to the justification of the present aspect of Guerlac's explanation. For what one needs to know is whether argument (A) was valid in the historical context in which Lavoisier found himself; that is, at best, we need to rely on the truth of (C) at the time of Lavoisier. To determine whether (C) was true one needs historical knowledge; thus the historian of science may be expected to be more qualified to judge that truth.

A similar argument could be given in connection with Koyré's and other explanations. Thus the main threat of Agassi's present argument, which attempts to make history of science logically dependent on the philosophy of science, is removed.

It should be noticed, however, that the historian does not even need the generalization that he would be competent to evaluate. In fact, all that Guerlac has to do, in order to justify his claim that assertions (1)–(4) explain the presumed fact that Lavoisier formulated the antiphlogistic conception, is to show that (1)–(4), if true, allow us to understand why Lavoisier formulated the antiphlogistic hypothesis. And one way to do this is to show that there is a comprehensible connection between the reasons mentioned in (1)–(4) and Lavoisier's conception. Evidence for the existence of such a connection would be provided by the presence in Lavoisier's reasoning of ideas adequate to link the four reasons to that conception. This is indeed what Guerlac tries to do. In fact, Guerlac's air-in-reduction-calcination thesis attempts unsuccessfully to establish the presence of such a link. Of course, even if Guerlac had established such a presence, the general connection could, in some sense, be questioned even further and thus require further justification. For example, one might now want evidence that Lavoisier could have adequately connected his antiphlogistic conception with the idea attributed to him in the above-mentioned thesis, namely that air is given off when metallic calxes are reduced and absorbed when metals are burned. And in my criticism of Guerlac I argued that Lavoisier probably could not have made this connection. But the fact that further justification may be required is not a special weakness of this way of justifying an explanation. In fact, it might be required even if we had produced the justification favored by Agassi. For example, one may question how (C1)–(C4) are to be combined with each other so as to yield Lavoisier's antiphlogistic conception from his four beliefs; or one might question the truth of any of those four generalizations.

For at least two different reasons, then, philosophy of science principles, of the kind conceived by Agassi, are not needed by the historian of science to justify his explanations. We also showed that those principles were not needed to find the explanations. History-of-science explanation is thus liberated from Agassi's restraints.

chapter 11

History of Science and Philosophy of Science

The Dependence of Agassian Philosophies of Science on History of Science

SUPPOSE THAT WE conceive of philosophy of science in the way Agassi does, namely, as the study and use of principles like inductivism, conventionalism, Popperian hypothetico-deductivism, and Archimedeanism, in other words, as theoretical history of science or philosophy of the history of science. Philosophy of science, so conceived, is logically and methodologically dependent on the historiography of science because, first, in order to find such principles, the philosopher of the history of science or theoretical historian needs history-of-science explanations. The two most common, if not the only ways of finding such principles are (1) to generalize a history-of-science explanation and (2) to generalize a personal experience. Suppose we know or believe that a certain scientist S succeeded because he used approach or method M, or that the logico-historical relation between a scientific theory and facts was that the theory was first hypothesized and factual consequences were later deduced from it. If the scientist is a great scientist or if the theory-hypothesis is an historically or contemporarily important theory, the tendency to generalize is obvious. The tendency may be reinforced by other justified or unjustified motivations. An example of the latter would be one stemming from the belief that same causes have same effects, which, together with the explanation that S succeeded because of method M, yields the conclusion that all scientific successes are due to the use of method M and all scientific failures or errors due to the misuse of M. An example of a justified motivation would be one stemming from the desire to determine if any other historical cases are like the one already explained.

I have already suggested the likelihood that Koyré formulated his Archimedean philosophy of science as a result of his explanation of Galileo's and Descartes's successes and failures concerning the discovery of the law of falling bodies. Another example is C. C. Gillispie's *The Edge of Objectivity*, the best example of an inductivist Archimedean philosophy of science. Much like Koyré, Gillispie emphasizes this methodological aspect of science at the same time that he uses an inductivist approach—inductivist in the sense that, for him as well as for Koyré, scientific knowledge consists of the contents of the latest textbook, which is the truth even though some of the statements contained therein are worse approximations than others, and no statement is a perfect approximation. Moreover, Gillispie's philosophy is probably the result of his accepting Koyré's explanation of Galileo's success. And this claim may be supported by two facts: (1) Gillispie begins with a discussion of Galileo's success along the lines of Koyré's explanation,[1] and he explicitly expresses his debt to Koyré.[2]

T. S. Kuhn's *Structure of Scientific Revolutions* is the best recent example of a conventionalist-methodological theoretical history of science. Kuhn, about whom more will be said later, also emphasizes the methodological or structural aspects of science. As regards the logic of science, Kuhn is a conventionalist; for him the intellectual content of science does not consist of things that can be either true or false. The principal source of Kuhn's philosophy is probably the

174

explanation of the Copernican revolution contained in his earlier book *The Copernican Revolution.*[3]

Popper and Agassi exemplify the hypothetico-deductive philosophy. Their work is at once an epistemology and a methodology of science; it claims that the relation between scientific theories and facts is that the latter are deducible from the former, which may thus be refutable but not provable by them; and it also claims that this conjecturing-refuting is the main feature of the scientific method. In other words, scientific theories are refutable or criticizable theories; and the scientific method is basically the critical attitude. It is the attitude that one adopts toward one's theories or ideas after he has them that makes one a scientist, and not the approach that one follows before or for the purpose of obtaining the theory. Regarding the source of the Popperian theoretical history of science, I believe that the Popperians themselves do not deny it to be the result of a generalization of their explanation of Einstein's successes.

So much for generalizing history-of-science explanations. I now turn to what I earlier called "generalizing from personal successes"— the method of arriving at philosophies of the history of science well illustrated by great scientists' formulation of their own philosophies of the history of science. The creative scientist who has made significant scientific contributions will have a certain view of what accounts for his success. He often generalizes the explanation of his own success into a philosophy of science. This, I think, is the way Newton and Einstein arrived at their philosophies.

Besides depending on history-of-science explanation for arriving at principles, the theoretical history-philosophy of science is dependent on it for the justification of those principles after they have been obtained and formulated. In fact, the application of such principles to historical cases *is* what Agassi conceives the philosophy of science to be, which is not surprising for I indicated before that, for Agassi, the philosophy of science is the philosophy of the history of science. And regardless of how we justify a philosophy of the history of science, whether by inductivism, conventionalism, or hypothetico-deductivism, an explanation of specific developments is needed. For example, Agassi would justify his Popperian philosophy of the history of science by arguing that it was the best explanation of the facts of the development of science. This argument will then be a series of explanations of various important and interesting history of science developments, i.e., a series of history-of-science explanations. Such explanations are then logically prior to the "philosophical" principles.

Analytical Philosophy of Science and History of Science Description

It follows, as an obvious consequence, that I do not deny, nor did Agassi claim, that all conceptions of the philosophy of science are relevant to or important for history-of-science explanation, still less to or for history of science in general. The question then arises: Is there a conception of the philosophy of science which is not dependent on history-of-science explanation in the above sense? The answer is yes. In fact, a second school of philosophy of science,[4] whose members may be called analytical philosophers, studies the nature of science by analyzing the logical structure of contemporary scientific knowledge. This analysis is carried out by inquiring into the nature of and logical interrelations among basic concepts which give scientific knowledge a structure. Some of these concepts are explanation, prediction, law of nature, theory, probability, definition, causation, confirmation.

It is obvious that analytical philosophy of science is independent of history of science; it is, however, dependent on the various sciences in a much more direct manner than is the philosophy of the history of science. I do not think that its independence results from the fact that analytical philosophy of science and historiography of science have nothing to do with each other. For though I do not find philosophy of science *analyses* (as we might call the analog of the "principles" of the philosopher of history of science) directly relevant to history-of-science explanations, they are indirectly relevant to them since they are directly relevant to history-of-science *descriptions*.

In Koyré's explanation, for example, we find him talking about Galileo's law of falling bodies and his definition of acceleration. Suppose we begin by asking Koyré why he describes Galileo's statement of the correct time-distance relation as a *law* and Galileo's principle of constant acceleration as a *definition*. Koyré might give the reason that he is simply describing those things as Galileo describes them. Koyré's reason would be inadequate: although Galileo does, at least in the *Discourse*, call his principle of constant acceleration a definition, he calls the law a *theorem*. Though inadequate, Koyré's reason points toward a distinction and raises a problem. The distinction is that between the description in the language of the historical agent, which may be called the *agent's description,* and that which the historian thinks is a correct description, which may be called the *historian's correct description.* The problem is what the relation between these two descriptions is or should be and whether the historian should stay away from the historian's correct description and limit himself to the agent's description. This problem may not have a general solution, and it may be unsolvable even in specific cases. (Consider the case of Priestley's discovery of dephlogisticated air. Did he discover such a non-existing entity, or did he think he discovered such an entity but "really" discovered oxygen? But did he discover modern [real] oxygen or did he only discover Lavoisier's oxygen, simultaneously with Lavoisier?) For our purposes here we need not, however, solve the problem. In fact, if instead of the historian's correct description, the historian adopts the agent's description, then what we could ask the historian, instead of asking him to justify his own descriptive claim, would be to tell us if the agent's description is correct. In this age of specialization some might be inclined to deny that the answer to such questions falls within the province of the historian himself, of the historian qua historian. But no one could deny that it would be desirable for the historian to answer those questions.

The next step is to determine how Koyré would support his claim that Galileo's statement of the time-distance relation is a law, really a law, a law in fact and not just in the mind of Galileo. As a reason Koyré might say that contemporary scientists call Galileo's time-distance relation a *law.* But for these scientists Galileo's time-distance relation is a certain formula appearing in their textbooks, and we trust that Koyré is not talking about such a formula, but about Galileo's own statement. Hence Koyré's reason would not do, but its reference to the present-day scientists represents a step in the right direction. For Galileo's statement is a law if it has the same logical characteristics possessed by statements which are normally and properly called laws by contemporary scientists.[5] Koyré, then, should know what these logical features are. If we accept Scriven's analysis, "we may say that typical physical laws express a relationship between quantities or a property of systems which is the *simplest useful approximation* to the true physical behavior and which appears to be *theoretically tractable.*" [6] "Theoretically tractable" here distinguishes between laws and theories. And Galileo did treat theoretically the time-distance relation. For as Koyré points out, in 1604 "Galileo is *already* in possession of such a formula . . . [but] he does not have any confidence in an observation which is not theoretically verified." [7]

If this is true, it raises a problem about another of Koyré's descriptions. Koyré claims that Galileo was the discoverer of the law, whereas it seems to be the case that Galileo only verified or proved the law. And what is the relation between Galileo's discovery and Beeckman's investigations? Why is not the latter a co-discoverer of the law? Clearly, in order to solve this problem and thus to justify the relevant descriptive claims, Koyré needs an analysis of the concept of discovery.

A third example may be taken from Guerlac's explanation. Guerlac describes Lavoisier's idea "that the air fixed in metals during calcination and released upon reduction might account for the greater weight of the calx" as "a more likely explanation" of the weight-augmentation effect. Guerlac's claim might, by a Baconian, for example, be challenged on the grounds that Lavoisier's idea is not an explanation of but a consequence of the augmentation effect. To defend his claim Guerlac would have to invoke principles which are part of the analysis of explanation. He might recede, of course, into describing Lavoisier's idea as a theory, a description, using a vaguer concept, which could be less easily criticized. However, that would still conflict with the claim someone might want to put forth that the idea is a consequence of a fact, the augmentation effect. For consequences of facts are also facts, and one would like to distinguish between facts and theories. Or Guerlac might say that he is merely describing the idea as Lavoisier would have described or perhaps does, in fact, describe it. This procedure that Guerlac would be following here might in particular cases be unobjectionable, but if adopted as a general escape route would be open to the objection that sometimes one wants to know not only what a historical agent says, but whether what he says is right.

My conclusion is that analytical philosophy of science and historiography of science are related as follows: philosophy of science analyses, of various structural concepts, are part of the grounds for history-of-science descriptions: analyses of the concept of law for descriptions involving that concept, analyses of the concept of explanation for descriptions involving the concept of explanation, and analyses of the concept of discovery for descriptions involving the concept of discovery.

Finally, it should be pointed out that although analytical descriptions are appreciated more by the analytical philosopher of science than by the philosophers of history of science, they are not entirely ignored by the latter, especially the Popperians, who have their own semi-technical concept: refutation. They in fact usually insist in describing, for example, Newton's theory of gravitation as a refutation of Kepler's laws and Einstein's theory as a refutation of Newton's.

chapter 12

Philosophy of the History of Science: Critique

History as the Study of Differences

THE PHILOSOPHY of the history of science is worse than "unnecessary for and dependent on" history-of-science explanation: the methodological legitimacy of the enterprise itself is in doubt. One reason for this may be connected with the practice and name of Koyré since he is much revered by the philosophers of the history of science no less than by simple historians of science. It is well known that Koyré likes to emphasize the differences rather than the similarities among various scientific thinkers, especially between a discoverer and his alleged or prima facie precursors. This is not at all an accident, for it is natural to be interested in the differences rather than the similarities in human behavior and events. (Hence arises a possible unbridgeable gap between the sciences and the humanities.) And insofar as the bulk of historiography studies the history of human affairs one could take the study of differences (irregularities?) to be one of the salient traits of the historical attitude. The study of the history of natural systems like the earth, the solar system, and the universe is either not counted as historiography but as part of the natural sciences, or if counted it constitutes a quantitatively small part of historiography. It is thus that a theoretical history of science, a philosophy of science in Agassi's sense, may be a contradiction in terms.

It is also a contradiction in practice. Philosophers of the history of science tell us that they are interested in the broad outline of the history of science (Agassi),[1] in the general patterns of the evolution of science, in the structure of all scientific change (Kuhn).[2] If their theories are to have any use, we might expect them to provide the explanation of fundamental and general facts of the developments of science. The most fundamental fact of that development is related to the existence and content of the latest textbook. The problem here is: How and why did that textbook become what it is; How and why did scientific knowledge develop into what it has; What is the origin of the latest textbook?

The Problem of the Selective Growth of Knowledge

The most striking large scale or sociological phenomenon in the whole history of the physical sciences, apart from the growth of knowledge itself, is that whereas Aristotelian and Cartesian physics, phlogistic chemistry, Newton's corpuscular optics, caloric thermodynamics, and the aether theory were abandoned or discarded when they were superseded, Galilean kinematics, Keplerian astronomy, wave-optics, phenomenological or macroscopic thermodynamics, Newtonian dynamics, pre-quantum-theory chemistry, and classical mechanics were retained when superseded. I shall refer to this fact as the selective growth of knowledge to draw attention to selectivity in the retention or abandonment of a previous scientific theory. The theories in the first group, like Cartesian physics, I shall term *invisible* because they cannot be found in any form in the latest textbooks. The theories in the second group, like Galilean kinematics, will be called *visible*. Similarly transitions from an invisible theory to a visible one, e.g., from Cartesian to Newtonian physics I shall call *invisible transitions;* whereas transitions from a visible to another visible theory, e.g., from Galileo's kinematics to Newton's dynamics, I shall call *visible transi-*

tions, since the student of science is even now required to make them at some stage of the process of his scientific education. The selective growth of knowledge must remain improperly examined as well as inexplicable, as long as it is not properly recognized, which it can hardly be, by the philosophers of the history of science, who are intent on searching for and finding a common structure to all scientific revolutions.

The Popperian Philosophy

The Non-Historical Character of the Conjecture-Refutation Description. According to the Popperians, knowledge grows by conjectures and refutations and all theoretical transitions are refutations of a previous conjecture by a new hypothesis.[3] The main difficulty is that the conjecture-refutation description, as it stands, is not a historical description at all; that is, it does not describe the theoretical transitions as they were envisioned by the historical agents. Instead it describes the situation as it appears to the historian of science or to the epistemologist; by contrast this might be called a theoretical description. This distinction between a theoretical and a historical description of a certain transition is not to be confused with the distinction often made between history and logic or between psychology and logic. The historical description will have, in fact, logical as well as psychological aspects. And in cases of scientific discoveries, revolutions, and the like, we can expect the logical aspects of the situation to be most important; but this logic, whether right or wrong, will have to be the logic of the historical agents. In other words, the logico-historical description of the situation will be the description of the situation from the logical point of view of the historical agents. Just as logical factors cannot be excluded from a historical description, so psychological and sociological aspects cannot be excluded from a theoretical description; a theoretical description has psychological and sociological as well as logical aspects.

To repeat: I claim that the conjecture-refutation structure is not a logico-historical structure, but a logico-theoretical structure; that is, it is not a structure in the mind of the historical agent but in the mind of the philosopher.

This contention is illustrated by the Popperians' discussion of Newton's alleged derivation of the law of gravitation from Kepler's laws. They do not claim, nor could they claim, that Newton believed that his law of gravitation refuted or was inconsistent with Kepler's laws. They admit that Newton did not think that his law of gravitation was a refutation of Kepler's laws. What they claim is that the law of gravitation is inconsistent with Kepler's laws, and consequently that Newton was wrong to the extent that he did not believe so. In other words, the law of gravitation is a refutation of Kepler's laws only logico-theoretically speaking, and not logico-historically speaking.

Let us agree, then, that the conjecture-refutation structure is intended to be the logico-theoretical structure of theoretical transitions. I can now show why this Popperian (logico-theoretical) description is objectionable. My objections are three: it is incomplete, misleading, and renders the phenomena described inexplicable.

The incompleteness of the description should be clear from the example of the transition from Keplerian astronomy to Newton's celestial mechanics. Even if the Popperians should be right about the logico-theoretical structure of all theoretical transitions, to say that they are refutations of earlier conjectures is not the whole story. We would like to know, when we are trying to describe the situation, not only that Newton's theory is a refutation of Kepler's conjectures, but also whether it was believed to be a refutation. The transition from Kepler to Newton is one example where the logico-theoretical structure, the "real" structure, does not coincide with the logico-historical structure. If it were an isolated case, then we might still regard the logico-theoretical structure as a good indication of the logico-historical one. But it is by no means an exceptional case; the transition from Galilean kinematics to Newtonian dynamics is another example of the dissimilarity of the two structures. A general argument why the two structures often diverge might run as follows. It might be supposed first that since men, including historical agents are, generally speaking, rational beings they might be right more often than not. To this one might reply that men are also fallible and will, therefore, often be wrong. Consequently these considerations are inconclusive. The decisive argument is grounded on the growth of knowledge—a fundamental fact. The growth of knowledge does, or at least can and should make the epistemologist, historian, and philosopher of the history of science wiser than the historical agents; so each will be able to discover logico-theoretical structures which either have no logico-historical counterpart, or if they do, are different from such counterparts. Thus the logico-theoretical structure of theoretical transitions will in general not coincide with the logico-historical structure, and to that extent, a logico-theoretical description is incomplete, since it provides no information about the historical situation.

My second objection to the Popperian description is that it is misleading, because, insofar as the Popperian is usually silent about the logico-historical structure of the transition, he insinuates that it is the same as the logico-theoretical structure which he does talk about. The insinuation is misleading because there is often no such similarity. On the other hand, if and when the Popperians are not silent about the logico-historical structure of the given transition, then the logico-theoretical description of the same transition is not really a description but an evaluation. For when we do know both the historical and the theoretical structure we can decide immediately whether the historical agents were right or wrong, that is, evaluate them. In fact, the only function played by the conjecture refutation is to provide an evaluation. There is no reason to disguise the evaluation as a description.

My third objection to the Popperian description is that it renders the occurrence of theoretical transitions inexplicable. The reasons are as follows. What we are trying to get at in describing the theoretical transitions of the selective-growth phenomenon are certain aspects of individual and collective human behavior. Now, it is a truism about human behavior that it depends on what human beings believe. Of course, their beliefs are not unrelated to reality so that there is a second-order dependence of human behavior on reality. Only secondarily, therefore, does human behavior depend on what really is the case, and then only through the correct beliefs held by individuals. If this is so, if human behavior depends primarily on what the individual concerned believes, then we can expect that historical transitions from one scientific theory to another can only be accounted for in terms of what the historical agents believed, that is, in terms of the historical structures involved. But if the selective growth of knowledge is described logico-theoretically—in terms of the structure which it "really" has—then it can only be accounted for logico-theoretically, i.e., in terms of some other real or theoretical structure. Thus, if the selective growth of knowledge is described logico-theoretically it cannot be explained or accounted for.

The Popperian Interpretation of the Selective Growth. Having described all theoretical transitions as refutations of past conjectures, the Popperians are forced to say that there is no important difference between the so-called invisible theories such as Cartesian physics and the visible ones such as Galilean kinematics. The reason given for this is that the so-called visible theories have really been discarded because it is only the formulas or the terminology which have been retained and not the meaning. In other words, they have been retained in appearance but discarded in substance.

As a logico-historical thesis, this interpretation is untenable. Consider, for example, the transition from Galilean kinematics to Newtonian dynamics, surely a visible one, since students are still required to make this transition in the early stages of their study of physics. Galileo's laws of falling bodies are mentioned by Newton at the beginning of the Scholium to the Axioms, or Laws of Motion,[4] in order to gain social and experimental support for the laws of motion that he has just stated (in particular to support the first two laws and first two corollaries). The support consists in the alleged fact that those laws can be derived from his axioms of motion. Admittedly, we can agree with the Popperian who would point out that Galileo's laws can be derived from the axioms of motion only as an approximation, so that if Newton believes otherwise he is wrong. But this is irrelevant for the present point; i.e., it was Newton's intention and belief that he could derive, perhaps approximately, some formulas or statements which not only looked like Galileo's laws but also were to be interpreted like them. It is obvious that he could not have believed both that Galileo's laws were consequences of his axioms and that these consequences were really something different from those laws, at least in substance. To attribute both beliefs to Newton is to attribute to him an inconsistency so flagrant that it could only reflect on whoever made the attribution. The beliefs concerning terminology and interpretation are in fact almost the opposite of what the Popperian theoretical description would lead one to believe. We find Newton actually stating Galileo's laws in his own terms; whereas Newton says that "Galileo discovered that the descent of bodies [i.e., of freely falling bodies] varies as the square of the time and that the motion of projectiles was in the curve of a parabola,"[5] Galileo first states the two laws as more or less mathematical theorems and then tries to show that the motions described by these theorems is the real motion of bodies. He states the theorems as follows: "The spaces described by a body falling from rest with a uniformly accelerated motion are to each as the squares of the time-intervals employed in traversing these distances,"[6] and "A projectile which is carried by a uniform horizontal motion compounded with a naturally accelerated vertical motion describes a path which is a semi-parabola." [7] Newton, then, is not so much worried about retaining Galileo's terminology as he is about retaining the meaning of the Galilean statements. It is the meaning of Galilean kinematics, then, that Newton believes he is retaining. To put it differently: there is every reason to believe that if Galileo had lived to see Newton's work he would have agreed with the latter's claims about his discoveries, that is, that he meant by his own statement of the laws what Newton meant by the statements attributed to Galileo. This Galileo would have claimed in spite of the many different theories with which the two men were working. In support of this hypothesis I would refer to the passage where Galileo[8] discusses the question of how one is to interpret his law about projectiles in view of the fact that the surface of the earth is spherical and thus the "horizontal" component of their motion is "really" an upward motion, a motion away from the center of the earth. Galileo makes it explicit that his law is an approximation. Thus we may conclude that if he had seen Newton's work he would have concluded that it contained the appropriate corrections to his law, which remained unaffected as an approximation.

On the other hand, if the Popperians' reinterpretation of the selective growth is regarded as a logico-theoretical thesis, then it generates the problem of what caused the selective retention of terminology, for which they seem to have no good answer. No doubt they would say that it was due to accidental, psychological, or unimportant factors. But this is tenable only as long as what is historical is equated or confused with what is accidental, or psychological, or epistemologically unimportant. As I indicated before, this identification is unwarranted. In fact, there are logical factors which are historical, if they refer to the logical beliefs of the historical agents. And it is probably to these logico-historical factors that the selective retention is due, as I shall argue in more detail later. If this is so, how could the consequences fail to be of epistemological importance? I suggest the primary epistemological consequence that, whereas the retained theories are in the domain of present scientific knowledge, the discarded theories are not only outside that domain but incommensurable with that knowledge.

Normative Aspects. The discussion of the epistemological consequences of the historical, i.e., logico-historical, factors producing the selective growth of knowledge leads to my criticism of the normative aspects of the Popperian philosophy. The main function that the logico-theoretical conjecture-refutation claims have in the Popperian philosophy of science is as grounds for the central tenet of their methodology. Feyerabend sums up this tenet as the "principle of proliferation," which he formulates as the demand to invent and elaborate theories inconsistent and incommensurable with the most highly confirmed prevailing theory.[9] Popper expresses it by saying that "the game of science is, in principle, without end. He who decides one day that scientific statements do not call for any further test, and that they can be regarded as finally verified, retires from the game."[10]

The Popperians do not, of course, try to derive or deduce the principle from the logico-theoretical thesis; on the contrary they would argue that one cannot derive what ought to be done from what is done. Moreover, they may have other grounds on which to base their demand. But the Popperians use the claim of a conjecture-refutation (logico-theoretical) structure to construct an argument in support of their methodological demand, and this argument is the only, or at any rate the most important, argument if their demand is to be regarded as a principle of the methodology of science, as contrasted to a moral principle of general ethics.

How, then, do the Popperians ground their demand for proliferation on the logico-theoretical claim? Their argument is an hypothetico-deductive one, so to speak. They argue that if their demand is reasonable, if their methodological principle is a good one, it would be expected to have played a decisive role at most, if not all, the major stages of the history of science, and perhaps of the history of thought in general. Their logico-theoretical thesis makes precisely this claim, namely, that this consequence of the methodological demand accords with the facts.

Now, I have pointed out previously that the facts with which the conjecture-refutation structure thesis accord are, at best, logico-theoretical, i.e., facts from the point of view of the philosopher. But the principle of proliferation cannot be grounded on such facts for it is a principle of action, and as such it must state the demand in a way which can be acted upon by the agent; it must state its demand in terms of the beliefs of the agent. In other words, the demand really means, or should be stated as follows: invent and elaborate theories which you believe are inconsistent and/or incommensurable with the most highly confirmed theory. One could question whether this demand is even self-consistent, but one thing is beyond doubt: it cannot be grounded on a logico-theoretical thesis but only on a logico-historical one. Since, as I have argued before, the theoretical and historical structures of theoretical transitions do not usually coincide, the demand for proliferation does not have a historical basis.

Popper's principle is similarly unfounded. All that he is entitled to say on the basis of the historical evidence is that, in principle, the game of science is without end. He who decides one day that his scientific statements do not call for any further test, and that they can be regarded as finally verified, may be entirely justified personally in his decision but can expect that sooner or later there will arise other conflicting scientific views which are an improvement on his own.

Criticism of Kuhn's Theory

For Kuhn[11] theoretical transitions are scientific revolutions; hence, to present his solution to the problem of the selective growth of knowledge, we shall have to state his doctrine of the structure of scientific revolutions. The doctrine has two main parts: a description and an explanation. The description consists of the account of those aspects of their structure which presumably allows us to establish their existence. I shall call this structural aspect of Kuhn's scientific revolutions their *identifying structure*. The explanation consists of Kuhn's account of how and why revolutions occur. I shall call this aspect the *genetic structure* of Kuhn's scientific revolutions.

Evaluative Aspects. One could wish that Kuhn's doctrine contained an evaluation of revolutions, but no coherent evaluative account can be abstracted. He renounces all evaluation in both word and deed. That is, he claims that it is impossible to judge whether, and if so to what extent, any revolutions were justified or unjustified, good or bad. Moreover, he never judges whether the Copernican, chemical, and Einsteinian revolutions, which he discusses, were good or bad. Typical of his nonevaluative mood, which is probably the prevailing one in his *Structure of Scientific Revolutions*, is the following:

> These characteristic shifts in the scientific community's conception of its legitimate problems and standards would have less significance to this essay's thesis if one could suppose that they always occurred from some methodologically lower to some higher type. . . . Yet the case for cumulative development of science's problems and standards is even harder to make than the case for cumulation of theories. [P. 108]

But Kuhn himself actively describes and engages in evaluation. He tells us in his introductory chapter that "the theses suggested . . . are . . . often interpretative and sometimes normative" (p. 8). And occasionally we find him actually involved in evaluation. For example:

> Priestley never accepted the oxygen theory, nor Lord Kelvin the electromagnetic theory, and so on. . . . These facts and others like them are too commonly known to need further emphasis. But they do need re-evaluation. In the past they have most often been taken to indicate that scientists, being only human, cannot always admit their errors. I would argue rather that in these matters neither proof nor error is at issue. The transfer of allegiance from paradigm to paradigm is a conversion experience that cannot be forced. Lifelong resistance, particularly from those whose productive careers have committed them to an older tradition of normal science, is not a violation of scientific standards but an index of the nature of science itself. [P. 151]

Finally, I suggest that Kuhn's whole theory presupposes a certain evaluation of the selective growth of knowledge, namely, that discarded theories are not unscientific (p. 2). In fact, he himself tells us that he aims to delineate a new image of science "by making explicit some of the new historiography's implications" (p. 3). The new historiography is, he claims, the result of two insights. The first is the difficulty and even impossibility of dating a discovery at an instant in time and of attributing it to one individual. The second is that there are two alternatives for the historian (p. 2): (1) "out-of-date beliefs are to be called myths" and "myths can be produced by the same sorts of methods and held for the same sorts of reasons that now lead to scientific knowledge," and (2) "they are to be called science," and "science has included bodies of belief quite incompatible with the ones we hold today." And "given these alternatives, the historian must choose the latter. Out-of-date theories are not in principle unscientific because they have been discarded" (pp. 2–3).

Besides being internally incoherent, the evaluative part of Kuhn's philosophy conflicts with the descriptive and explanatory parts. In fact, as I shall show, his description of revolutions entails that evaluation of them is possible, and his explanation of their occurrence entails that they are to be evaluated favorably.

The Definition of Scientific Revolutions. Kuhn holds that "scientific revolutions are . . . taken to be those non-cumulative developmental episodes in which an older paradigm is replaced in whole or in part by an incompatible new one" (p. 92).

Notice that Kuhn does not speak of a theory but of a paradigm. A paradigm is a sociological entity and refers to the scientific community's set of professional commitments of various kinds, instrumental and methodological as well as conceptual and theoretical. In this way experimental, observational, and theoretical discoveries can all be treated under the heading of "paradigm innovations," a notion that serves well someone who, like Kuhn, believes that "scientific fact and theory are not categorically separable, except perhaps within a single tradition of normal-scientific practice" (p. 7).

Second, a paradigm innovation is not yet a scientific revolution since the latter is a paradigm change of a certain kind; the innovation, therefore, must be "assimilated" on the part of the scientific community before we have a revolution (p. 95).

Third, Kuhn's description speaks of a paradigm replacement "in whole or in part," which allows for the possibility of both major and minor revolutions. These correspond to large or small paradigm changes respectively, anywhere from the "major turning points in scientific development associated with the names of Copernicus, Newton, Lavoisier, and Einstein" (p. 6), to the somewhat smaller paradigm changes produced by the wave theory of light, the dynamical theory of heat, and Maxwell's electromagnetic theory (p. 66), down to the minor paradigm change produced by Hershel's discovery of Uranus (p. 116).

Fourth, in describing the development as "non-cumulative" Kuhn refers to the differences of problems as well as of problem solutions between the two paradigms; this can be expressed by saying that the new paradigm, besides being incompatible, is incommensurable with the old. A scientific revolution is then the transition to a new, incompatible, and incommensurable paradigm (p. 148).

Now, it is obvious that Kuhn cannot mean that the post-revolutionary paradigm is new, incompatible, and incommensurable from the same point of view. In fact, nothing, let alone a paradigm, can be both incommensurable with something else and also incompatible with it and new. If the two paradigms are incommensurable then one cannot compare them and conclude that they are incompatible and that one is conceptually old and the other new. And conversely, if we can say that they are incompatible and one old and the other new, then to that extent the two paradigms are commensurable.

Forgetting for the moment about novelty, what Kuhn does mean, and what he must, is that the incommensurability pertains from one point of view and the incompatibility from another. In fact, if we recall Kuhn's arguments to support his incommensurability claims we realize that they are aimed to show, and can only show, the incommensurability of the two paradigms from the point of view of either paradigm, and not their objective incommensurability from our point of view (the historian's paradigm). We might call the former *mutual incommensurability* and contrast it with the latter which might be called *real* or *absolute* (or *unqualified*) *incommensurability*. (Note, however, that both are objective in the sense of inter-subjective; they are not relativized with respect to individuals.) But, whenever Kuhn tries to support his incompatibility claims, he does not relativize the discussion to any paradigm (though, of course, he is speaking from our, the historian's, point of view). The incompatibility is then absolute, and we may speak of it without qualification. A revolution is then a transition to an incompatible and mutually incommensurable paradigm.

But if this is so, if the historian can compare and is comparing the pre- and post-revolutionary paradigms and concluding that they are mutually incommensurable, then Kuhn's refusal to evaluate revolutions is seen to be groundless.

This brings me to the subject of novelty. The post-revolutionary paradigm is "new" according to Kuhn. This novelty is not just temporal; the post-revolutionary paradigm is new not only in the sense of being more recent or younger or later. The novelty is primarily cognitive, for it is obvious that the paradigm is the scientific community's set of theoretical, conceptual, methodological, and instrumental commitments. It is also obvious that Kuhn contrasts transitions to a new paradigm, or paradigm changes, that is, scientific revolutions, to paradigm elaborations or normal scientific developments. But what is the point of view from which the post-revolutionary paradigm is new? For reasons analogous to those used in determining incompatibility, and also for the reason that paradigm novelty is a consequence of paradigm incompatibility, the novelty is unqualified, that is, not relative to any point of view.

Criticism of the Definition. To be historically real, to be a historical episode, the transition in question must be a transition to a paradigm believed by the historical agents to be new, incompatible, and mutually incommensurable. Whether or not the successful paradigm is new, incompatible, and mutually incommensurable, a revolution can be expected to take place only if the historical agents make a transition to what they believe to be new, incompatible, and mutually incommensurable paradigms. Of course, since men, including the historical agents, are, generally speaking, rational beings, they will be right more often than not. Because men are also fallible, however, and because of the fundamental fact of the growth of knowledge, which does or at least can or should make the historian wiser than the historical agents, the latter will not always be right. Therefore, (1) transitions to new, incompatible, and mutually incommensurable paradigms do not always coincide with (2) transitions to paradigms believed by the historical agents to be new, incompatible, and mutually incommensurable. The consequences of this fact will be examined shortly. But let us first ask whether Kuhn recognized it as a fact.

That Kuhn never makes the distinction between (1) and (2) is obvious from the text. That he actually proceeds as if there were no such distinction is shown, for example, by his argument for the mutual incommensurability of Newtonian and relativistic mechanics: he cites no historical evidence which might show that the historical agents believed or believe that the two paradigms are incommensurable; rather Kuhn tries to show that they are incommensurable.

There are also independent reasons to suppose that Kuhn's identifying structure is not a historical structure. For if the identifying structure were a historical structure then a Kuhnian scientific revolution would be a transition to a paradigm believed by the historical agents to be new, incompatible, and mutually incommensurable. This may be called Kuhn's *intended* identifying structure.

The first question to be asked about this formulation is whether it is in general possible to believe that a successful or later paradigm is new and incompatible and mutually incommensurable. I think it is probably a psychological impossibility to believe all this. For if one believes that the later paradigm is incompatible with the earlier, then if one has discovered this incompatibility himself one will certainly believe that the two are mutually commensurable; whereas if one accepts the incompatibility on other grounds, he *may* not believe in the mutual commensurability, but he almost certainly will not believe in the mutual incommensurability. By contrast, if one believes that the two paradigms are mutually incommensurable, he will at most believe that they provide different descriptions but probably not that they are incompatible. He hardly could have discovered this incompatibility without comparing the two paradigms. This point should not be confused with my earlier discussion of the consistency of the set of attributes of novelty, incompatibility, and mutual incommensurability. I argued that these attributes are consistent since the incompatibility and incommensurability held from different points of view: the incompatibility was unqualified (i.e., absolute), the incommensurability mutual. Now I am arguing that the set of attributes of believed novelty, believed incompatibility, and believed mutual incommensurability is inconsistent since the historical agents, necessarily both as historical and as agents, take only one point of view, their own. In other words, what is being here examined is a question of historical belief, what I examined earlier was a question of being. I established that there could be transitions to new, incompatible, and mutually incommensurable paradigms, since the notion is not contradictory. What I am trying to establish now is that there cannot be transitions to paradigms believed by the historical agents to be new, incompatible, and mutually incommensurable since this notion is logically or empirically self-contradictory. The Kuhnian scientific revolutions are indeed theoretical fictions rather than historical episodes.

But what about Kuhn's examples? Do they refute my preceding argument? Kuhn refers to many "revolutions," but he discusses in some detail only three: Copernicus's, Lavoisier's, and Einstein's (relativity). It is clear that neither Copernicus nor Galileo, nor any of the other participants in the debate believed that Copernican astronomy was mutually incommensurable with Ptolemaic astronomy. There would have been no such discussions if the historical agents believed the two theories to be mutually incommensurable. Therefore, despite their belief that the heliocentric system was conceptually new and incompatible with the geocentric, the Copernican revolution itself does not have the intended identifying structure. Kuhn might argue that though this intended structure is absent the revolution does have the theoretical identifying structure. But besides being historically irrelevant, this is disputable. For if Kuhn is willing to say that someone who refused to believe in Copernicanism must have thought that it was mad to claim that the earth moves, since what he meant by "earth" was the fixed point of reference of everything; if Kuhn is willing to accept as much, then why not argue that the two paradigms were in fact equivalent, because of the relativity of motion, etc., and hence not even incompatible.

As for Lavoisier's revolution, he must indeed have believed that his oxygen theory of combustion was incommensurable with the phlogiston theory; in fact, he made sure that this was so by building a "new system of chemical nomenclature." And certainly he managed to convince most of the scientific community of this. But this does not mean that Lavoisier's revolution had the intended identifying structure. He must have believed that the two paradigms were not wholly incompatible, for he was quick to recognize what Priestley's "dephlogisticated air" or Cavendish's "pure phlogiston" were or could be in the new system. Lavoisier surely recognized that, though partly incompatible, the two paradigms were often different descriptions of the same things and in that sense compatible. Nor does Lavoisier's revolution have the theoretical identifying structure since, in all likelihood, this structure happens in this case to coincide with the actual historical one.

As for the structure of Einstein's relativity revolution, all that Kuhn tries to show is the mutual incommensurability of relativistic and Newtonian mechanics. He does not even try to show that the two paradigms are incompatible. Nor does Kuhn try to show that the two paradigms were or are believed by scientists to be incompatible and mutually incommensurable. Therefore, Kuhn has demonstrated neither that Einstein's relativity revolution has the intended historical identifying structure nor that it has the theoretical identifying structure.

Kuhn's main examples thus do not fit his own description either when it is interpreted as stated, as a theoretical structure, or when it is interpreted as intended, as a historical structure. In fact, our discussion of Kuhn's examples shows that novelty, incompatibility, and mutual incommensurability, believed or real, do not necessarily occur together. This we also earlier discovered from a general consideration of those attributes. Thus Kuhn's description of "scientific revolutions" fails fundamentally in one of its primary aims, namely, to allow us to establish their existence.

The Occurrence of Revolutions. But let us see what Kuhn says about the causes of the occurrence of his revolutions; perhaps we can find there some of the history wanting in their description:

> the revolutionary competition between proponents of the old normal-scientific tradition and the adherents of the new one . . . [is] the process that should somehow, in a theory of sci-

entific inquiry, replace the confirmation or falsification procedures made familiar by our usual image of science. [This c]ompetition between segments of the scientific community is the only historical process that ever actually results in the rejection of one previously accepted theory or in the adoption of another. [P. 8]

What is the process by which a new candidate for paradigm replaces its predecessor? . . . What causes the group to abandon one tradition of normal research in favor of another? [P. 144]

Just because it is a transition between incommensurables, the transition between competing paradigms cannot be made a step at a time, forced by logic and neutral experience. Like the gestalt switch, it must occur all at once (though not necessarily in an instant) or not at all.

How, then, are scientists brought to make this transposition? Part of the answer is that they are very often not. [P. 150]

Still to say that resistance is inevitable and legitimate, that paradigm change cannot be justified by proof, is not to say that no arguments are relevant or that scientists cannot be persuaded to change their minds. Though a generation is sometimes required to effect the change scientific communities have again and again been converted to new paradigms. . . . We must therefore ask how conversion is induced and how resisted.

What sort of answer to that question may we expect? . . . when asked about persuasion rather than proof, the question of the nature of scientific argument has no single or uniform answer. [P. 152]

Ultimately, therefore, we must learn to ask this question differently. Our concern will not then be with the arguments that in fact convert one or another individual but rather with the sort of community that always sooner or later forms as a single group. [P. 153]

Kuhn[12] is determined to find a historical "structure" of scientific revolutions. He rightly feels, therefore, that he must exclude the logician's structure. He thinks he has excluded it by excluding the logical considerations which affected the historical agents. But such considerations would provide what may be called the logico-historical structure of revolutions and cannot be excluded in historical studies. What he should have excluded is the unqualified and confused talk about incompatibility and incommensurability; for as it is obvious from our previous discussion, these remarks have all the appearances of talk about the logician's structure. Kuhn's first mistake, then, is to abandon the logic of the historical agents.

Having summarily dismissed the historical agent's procedures of falsification and confirmation, he goes on to discuss the relevant techniques of persuasion. Finding no universals here either, he moves from psychology to sociology.

Here he concludes that he has found some regularity. Yet he has found it only because he has chosen to leave history behind. In fact, he fails to discuss the only historically relevant question: Is the post-revolutionary scientific community numerically identical with the pre-revolutionary scientific community. Had he paid attention to this question he would have had to recognize that the pre- and post-revolutionary communities are sometimes numerically identical, sometimes numerically different. I suggest that in the case of the invisible transitions two distinct communities were actually involved, whereas only one was involved in the case of the visible transitions.

Solution to the Problem of the Selective Growth

To start with the simplest case, it is well known that Newton accepted Galileo's kinematics and Kepler's astronomy, but rejected Cartesian physics. Now for both Kuhn and the Popperians the transitions from Galilean kinematics to Newtonian dynamics, from Keplerian astronomy to Newtonian celestial mechanics, and from Cartesian physics to Newtonian physics are qualitatively similar. They are all alleged to be "scientific revolutions," "paradigm changes" (Kuhn's terminology), or "refutations of earlier conjectures" (Popperian terminology). In my view the only revolution was the transition from Cartesian to Newtonian physics; the other two transitions were important evolutionary changes, but not revolutionary ones.

As we know, Newton rightly or wrongly accepted Kepler's astronomy and Galileo's kinematics but rejected Cartesian physics. His acceptance of Kepler's astronomy consisted in the belief (*mistaken* belief, we shall be reminded by Kuhn and the Popperians, but *actual* belief nevertheless) that he had derived his law of gravitation from Kepler's laws. Nor can it be disputed that he used Kepler's laws in the alleged derivation. Newton believed, then, that he was improving Kepler's results. The acceptance of Galileo's kinematics consisted of a similar belief. In other words, Newton believed Kepler's and Galileo's results to be commensurable and compatible with his own views.

In what did Newton's rejection of Cartesian physics consist? As Koyré has unwittingly shown[13] Newton's disagreements with Descartes were numerous and fundamental. Newton must have felt, then, that he could not "repair" the Cartesian system, or even if it could be repaired it was not worth repairing; it was better to build a new system. He must have believed then, that Cartesian physics was either incompatible or incommensurable with his own views.

Although the difference between the three transitions is easily recognized and explained, the evaluation of the facts constituting them is much more difficult. It is well known that Newton was wrong in his belief that he had derived his law of gravitation from Kepler's laws. In fact, Newton's mechanics is, strictly speaking, incompatible with Kepler's astronomy and Galileo's kinematics. Yes, "strictly" speaking; but not "approximately" speaking; for Kepler's and Galileo's laws can be derived as approximations within Newton's mechanics. At any rate their incommensurability has never been, and I can venture to say, cannot, be shown. Therefore, Newton would be right in his basic belief that he was improving on Galileo's and Kepler's results.

As for Newton's belief that his theories were certainly incompatible or incommensurable with Cartesian physics, anyone who has tried to understand Cartesian physics must reach the conclusion that the two theories are, by and large, incommensurable. Insofar as they are commensurable, they are frequently incompatible. Once again Newton is right in his judgment of the relation between his mechanics and Cartesian physics.

Although my evaluation of Newton may be contestable, it is important to note what the effects of this and any other evaluation of Newton are or could be. No evaluation could change the historical facts of the case; it would at best relate them to other cases.

It cannot be denied that the structure of scientific change, the historical structure (which does *not* exclude all logical or sociological structure, but only our evaluations since it can and indeed must include the logico-historical facts, i.e., the logic of the historical agents), primarily depends on the beliefs (scientific and logical, right or wrong) of the historical agents. Now I conjecture that the (historical) difference between the discarded and the retained theories springs from reasons similar to those in Newton's case. Those theories are discarded which are believed rightly or wrongly by the innovators to be perhaps incommensurable but at any rate largely incompatible with the new theory. Those theories are retained which are believed, perhaps wrongly, to be susceptible of improvement. Of course, the only proper, or historically real, revolutions are the former; the latter can at best be "logician's revolutions."

If I am right in the above conjecture, there can be no minor historically real scientific revolutions, for insofar as the change is a minor one it will be possible to regard it as an improvement and it will probably be so regarded by the historical agents. This will allow for a scientific development in which later theories become more and more, "strictly speaking," incompatible and even mutually incommensurable with the much earlier ones. The incompatibility and (or) mutual incommensurability is, however, of no concern, since the transition not only can be made but was historically made in small steps, each of which was regarded as an improvement.

If and when exceptional individuals can and do find that they can themselves make numerous "improvements," so numerous in fact that the result is hardly commensurable with the criticized theory, they will probably be inclined to feel that the prevailing theory is not worth "repairing," that it is better to build a new one instead. In this, of course, they will be right since their limited lifespan would not be sufficient to bring to a conclusion the controversies connected with the direct criticism of a prevailing theory. According to my above conjecture, the individual will reject the prevailing theory, and if he is right, so will the resulting scientific community. The old community will simply die out. My conjecture, then, can account for the fact that major revolutions, proper or historically real ones, are often correlated with one individual and with an old scientific community which "dies out."

PART THREE

Understanding the History of Science

chapter **13**

Scholarship, History, and Chronicle

IF THE NATURE of the deficiency of history-of-science explanation is not that wished by Agassi, neither is the nature of the inadequacy of Agassi's position that imagined by historians of science. They have objected to his criticism by saying that it is largely irrelevant, because it is grounded on the examination of works not written by members of their profession, and because it ignores that the study of the history of science has now become professionalized.

An exponent of the first argument is Charles C. Gillispie who alleges that in "castigating what he calls the 'inductivist' school, Agassi . . . adduces as representative historians of science . . . (to name some whose feelings will no longer be involved) Laue, Cajori, Jeans, Roscoe, White, Pledge, Wolf, Dampier-Whetham, and Laplace. What Agassi is really criticizing, therefore, is the conception of science which has animated many scientists, but surely that is a larger matter to be handled relative to science itself rather than to the attempts at history to which some have turned, often in advanced years, and which many of us would prefer to let be as the work of amateurs, antiquarians, or chroniclers." [1]

Analogously, Thomas Kuhn argues that Agassi's "tone suggests that a principal aim was reform in the contemporary professional practice of history of science. But for this task his monograph is dreadfully misdesigned. To castigate contemporary practitioners by analyzing the historiography of James Jeans . . . H. E. Roscoe . . . or Max von Laue . . . is laughable." [2]

But scientists' histories of science should not be dismissed as the work of amateurs, antiquarians, or chroniclers, and the analysis of them is not laughably irrelevant to the historiography of science, an endeavor whose scope may be defined by making appropriate adaptations in the concept of 'history' formulated by Benedetto Croce:

> What constitutes history may be thus described: it is the act of comprehending and understanding induced by the requirements of practical life. [In some situations] these requirements cannot be satisfied by recourse to action unless first of all the phantoms and doubts and shadows by which one is beset have been dispelled through the statement and resolution of a problem—that is to say—by an act of thought. In [this] seriousness of some requirement of practical life lies the necessary condition for this effort. It may be a moral requirement, the requirement of understanding one's situation in order that inspiration and action and the good life may follow upon this. It may be a merely economic requirement, that of discernment of one's own advantage. It may be an aesthetic requirement, like that of getting clear the meaning of a word, or an allusion, or a state of mind, in order fully to grasp and enjoy a poem; or again an intellectual requirement like that of solving a scientific question by correcting and amplifying information about its terms through lack of which one had been perplexed and doubtful. . . . Historical works of all times and of all people have come to birth in this manner and always will be born like this, out of fresh requirements which arise, and out of the perplexities involved in these. [3]

If Croce's concept of history is accepted, the historiography of scientists' histories will be highly relevant, and neither laughable nor dismissable, to anyone interested in the history of science, including, presumably, professional historians of science. The reason is that, for a scientist, the requirements of practical life mentioned by Croce are the necessities of his scientific investigations. And the perplexities involved in them, which for Croce would provide the logical motivation for a historical work, are scientific predicaments whose resolution is aided by studying the development of science. Clearly, scientists are the category of people most likely to face such perplexities and hence to engage in those acts of comprehension and understanding which Croce regards as historical work and which from the nature of the case would constitute a history of science. Priestley's *History and Present State of Discoveries Relating to Vision, Light, and Colours* and Einstein and Infeld's *The Evolution of Physics* are foremost examples of this concern. But to a lesser degree, Jeans, Roscoe, von Laue, and the others named by Gillispie and Kuhn, were probably facing the same type of doubt, whose formulation and solution constitute their histories of science.

The second objection by historians of science to Agassi's criticism is that it ignores the recent professionalization of the study of the history of science. Roger Hahn, for example, claims that "as the history of science has fallen into the hands of a gradually organized profession, the blatant motivations—of any type—which colored past historical writings have slowly receded. By a constant and critical confrontation of opposing views among specialists, the purpose of examining the history of science has slowly been transformed. The older external motivations have given way to an interest in the subject for its own sake and a gradual transformation has been in the making for the last few decades. As happens in other human endeavors, professionalization has already raised the standards in the discipline and there is every reason to think that it will continue to do so in the future." [4]

He wants to argue that professionalization must have eliminated unsatisfactory work in the discipline because professionalization eliminates external motivations and establishes an interest in a subject for its own sake. The lack of external motivation probably does raise the standards of scholarship, but whether this improves the discipline, history of science, is another matter, a problem which has already arisen in another form in our discussion of the first objection to Agassi's criticism by historians of science. There the question was whether scientists, though nonprofessional historians of science, could write good histories of science. Hahn's view forces us to ask a somewhat complementary question, whether professional historians of science necessarily write

good histories or better histories than scientists or philosophers who
have "external" motivations to write them.

External motivations are indispensable for the writing of history if,
again, by 'history' we use Croce's definition: "the act of comprehending
and understanding induced by the requirements of practical life. . . .
[which] cannot be satisfied by recourse to action unless first of all the
phantoms and doubts and shadows by which one is beset have been
dispelled through the statement and resolution of a problem—that is
to say—by an act of thought." According to Croce's concept of history,
the elimination of external motivation eliminates the possibility of history. Consequently, far from being better qualified to write history, the
professional scholar is, qua scholar, incapable of writing history. To be
exact, the scholar is necessarily so limited—unless he explicitly and
consciously adopts and practices a logical style which may be called
"explanatory style," and roughly characterized as "problem-oriented
exposition."

The role of the scholar, however, is necessary, useful, and important.
History could not be started without his narratives and documents.
He is doing on a larger scale what, in Croce's words, "we all of us do
at every moment when we note dates and other matters concerning our
private affairs (chronicles) in our notebooks," [5] or "when we tie a knot
in our handkerchief in order to be reminded of something." [6] Yet, as
Croce also says, "in theory and in fact the two things are separate, and
neither the dull metal of the chronicles nor the highly polished metal
of the philologists [i.e., scholars] will ever be of equal value with the
gold of the historian even if that is concealed in dross." [7]

These consequences we have to accept if we accept Croce's concept
of history, which I think we must do for two reasons. First, Croce's
theory of history actually corresponds to the concept to which everyone
pays lip service. According to this concept a history is a series of causal
statements about the past, and it is contrasted to a chronicle which contains descriptions of facts without their causal connections. The same
idea is sometimes conveyed by saying that a history must be an interpretation and not just a description of past facts. This notion takes what
is perhaps its most relevant and sophisticated form when it is said that
a history must consist of explanations. For example, Morton White
says that "since a history makes reference to explanatory connections,
we may conceive of a history as a logical conjunction of explanatory
assertions." [8] It is indeed true that histories must consist of explanatory

assertions, but such assertions are, in general, neither 'because' statements, as White seems to think, nor answers to 'why' questions, as other historians and philosophers think. Rather, at the most general level, explanation is the reduction of what is *not* understood to what *is* understood, and explanations are whatever provide understanding when it is lacking, as Michael Scriven has emphasized.[9] If a history is a series of explanations, then the reading or the writing of a history would be a series of acts of comprehension and understanding, of which Croce speaks. And just as an explanation is not primarily a linguistic entity, but rather the act, often the speech act, which provides understanding, so a history for Croce is not primarily a book but a series of psychological or mental acts of understanding. This is why he so often makes fun of scholars, who think they can keep their histories locked in a library. They make the same mistake as do the positivist formalist philosophers of science when they define an explanation as a set of sentences of appropriate form and relation, which can be written down on paper and pointed to.

The second reason for accepting Croce's concept of history is that it is the least problematic way of making the distinction between history and chronicle, which any concept of history must do. All other attempts to differentiate between the two run into difficulty because they give formal or structural criteria for the distinction, that is, they characterize the difference in terms of properties possessed by one and not the other, e.g., how well interconnected are the facts contained therein or what is the order in which the events are treated. Such criteria are not only beset by the problem of the continuum between causal and noncausal language,[10] but also by the existence of countless cases of less well organized works which are better histories than better organized ones, and of better organized ones which are worse histories than more poorly organized ones. A way of solving all these difficulties is to adopt a functional and psychological criterion. Thus, the difference between history and chronicle becomes a difference in the "function" that the writing or the reading of a given work performs for the relevant individual—which individual is relevant depending on the content. History enables one to solve the theoretical problems arising from and blocking one's behavior; chronicle does not. History is not then something *more than* chronicle. The relation between the two is similar to that between description and explanation, as conceived by Scriven.[11] In Croce's words:

The truth is that chronicle and history are not distinguishable as two forms of history, mutually complementary, or as one subordinate to the other, but as two different mental attitudes. History is living chronicle, chronicle is dead history; history is contemporary history, chronicle is past history; history is principally an act of thought, chronicle an act of will. Every history becomes chronicle when it is no longer thought, but only recorded in abstract words, which were once upon a time concrete and expressive. Even the history of philosophy is chronicle when written or read by those who do not understand philosophy: history would be even what we are now disposed to read as chronicle, as when, for instance, the monk of Monte Cassino notes: 1001. *Beatus Dominicus migravit ad Christum.* 1002. *Hoc anno venerunt Saraceni super Capuam.* 1004. *Terremotus ingens hunc montem exagitavit,* etc. [1001. Holy Dominic leaps up to Christ. 1002. This year the Saracens came onto Capua. 1004. A big earthquake has shaken this mountain]; for those facts were present to him when he wept over the death of the departed Dominic, or was terrified by the natural scourges that convulsed his native land, seeing the hand of God in that succession of events. This does not prevent that history from assuming the form of chronicle when that same monk of Monte Cassino wrote down cold formulas, without representing to himself or thinking their content, with the sole intention of not allowing those memories to be lost and of handing them down to those who should inhabit Monte Cassino after him.[12]

I conclude that external motivations are indispensable to the historian of science also. A lack of them cannot result in the raising of history-of-science standards. This conclusion defines a sense in which Agassi's position is valuable. The concept of history he presupposes, is valid; it is Croce's definition applied to the history of science.

chapter 14

Explanation as Real History

THE PRECEDING CRITICISM of history-of-science explanations uncovered several deficiencies in explanatory practice. I do not claim, nor was it my intention to show, that there are no satisfactory aspects of that practice. Still less do I claim to have shown that *all* explanatory practice or all historical practice in general is unsatisfactory. Perhaps it is already obvious that the three explanations examined are not unsatisfactory in the same ways (i.e., for analogous reasons). The possibility thus arises that some of them are satisfactory, even highly valuable, in ways in which some other explanation is not.

To distinguish their positive from their negative aspects, and eventually find the cause of their deficiencies, demands a fuller logical analysis than given previously. In fact, a likely source of failure in Agassi's account is carelessness about the philosophical analysis of explanation related to his uncritical acceptance of Popper's "deductive model."

History as Explanation

Whatever their deficiencies, the one great merit of the works criticized earlier is that they are highly readable, even engrossing. In this respect they are in the same category as T. S. Kuhn's *The Copernican Revolution,* E. A. Burtt's *The Metaphysical Foundations of Modern Physical Science,* and A. Koestler's *The Sleepwalkers.* And they contrast sharply with a book like E. J. Dijksterhuis's *The Mechanization of the World Picture,* which is by far the most scholarly, accurate, and competent work on the history of science with which I am acquainted. Yet few who have tried to read it will deny that they find it dull. This contrast needs to be mentioned only because it illustrates certain points about (1) explanation, (2) historiography of science, (3) scholarship, and (4) history in general.

First, works like Guerlac's and Koyré's engage our attention because they are or contain explanations. I mentioned that the authors explicitly put them forth as explanations. This is of course no proof that those works actually *are;* it is at best proof that Guerlac and Koyré are aware of what logical activity they are engaged in. Why are they right in saying that their main purpose is explanatory? What features are possessed by the logical activity in which they are engaged by virtue of which it is an explanation? Certainly it would not do to say with Popper that "to give a *causal explanation* of an event means to deduce a statement which describes it, using as premises of the deduction one or more *universal laws,* together with certain singular statements, the *initial conditions.*" [1] Koyré and Guerlac do not deduce descriptions of Galileo's and Lavoisier's discoveries from universal laws and initial conditions. And only in special contexts does explanation require that kind of deduction. The most basic and fundamental logical feature of what Guerlac and Koyré do is to attempt the reduction of something which is not understood to what is understood; nothing more and nothing less is required for explanation than to provide understanding when it is lacking. [2] Each begins by formulating a problem, and most of the rest of each book attempts to solve that problem, along with smaller ones encountered in the course of that solution. The formulation of a problem is a statement that something is not understood, while the solution is the material or process making it comprehensible. For example, Guerlac's problem—what is not understood about Lavoisier's development in 1772—is that Lavoisier in the fall of 1772 discovered the explanation of the augmentation effect, allegedly as a result of certain experiments he had performed, even though as late as the beginning of that year Lavoisier had shown no interest in calcination, combustion, and the chemical role of air. The problem, then, is that the origin of Lavoisier's discovery is not known, and not knowing that origin we do not understand how Lavoisier made or began to make his discovery. What Guerlac attempts to do in his solution is to provide that origin, thus making Lavoisier's behavior and thinking comprehensible. As Guerlac narrates his solution he encounters other problems of understanding and hence provides other explanations. One such problem, as Guerlac sees it, is that although Lavoisier had conceived of the antiphlogistic explanation as early as August 8, he did not perform the experiments until late October.

Similarly, Koyré's problem—what is not understood about the discovery of the law of falling bodies—is that both Descartes and Galileo made various errors in the process of discovering it, even though the law appears to us very simple and can easily be grasped by beginners nowadays. If one could find in Descartes's and Galileo's mental attitudes some features—partly natural and partly idiosyncratic—which could account for their errors, then the errors would become understandable. And as Koyré tries to make them understandable, he encounters other problems to be clarified, such as Beeckman's correct derivation in spite of Descartes's error, and Galileo's own final success in spite of his own earlier errors. In turn Koyré tries to make these comprehensible.

We are now in a position to account for the interest and readability of Guerlac's and Koyré's works on two bases: first, they are or contain explanations; second, explanation is the reduction of what is not understood to what is understood. On this account of what an explanation is, we must have in an explanatory context (1) that which is not understood, (2) that which is understood, and (3) the reduction from one to the other. What is not comprehended is normally called, or formulated in, a problem. The initial formulation of the problem arouses our interest. This is not a psychological accident; a problem is not a problem if one does not feel intellectually (logically, if you will) a lack of understanding. And feeling a lack of understanding is perceiving logical conflict that demands clarifying. But the perception of logical conflicts necessarily carries with it, as a kind of logico-psychological concomitant, the desire to eliminate the conflict, hence the desire to read on in the work which has begun by formulating a problem. If in the working out of the initial or main problem, other uncomprehended things emerge and are in turn explained, as in the case of the two works being considered here, then it is obvious that additional logico-psychological motivations are provided; the work being read then continues to be, or becomes more, absorbing.

I contrasted above the interest and readability of works like Guerlac's and Koyré's to the dullness and unreadability of books like Dijksterhuis's. That works like the former are the exception rather than the rule has been noted by Agassi.[3]

In a different context, and in terms of relevance, a similar phenomenon has been described by Kuhn. In discussing the relevance of the history of science, he admits that the history of science will probably prove to be least relevant to the professional scientist.[4] Let us remember now that the scientific community is to history of science what intelligent and literate laymen are to socio-political history, so that scientists would seem to be the most natural readers-consumers of historiography of science. Let us also remember that to speak of relevance is another way of referring to interest and readability. We then see that Kuhn is talking about the same phenomenon as Agassi.

Finally, I. B. Cohen noticed the same characteristic when he told his colleagues not to worry about "bridging the gap" among disciplines, such as the sciences and the humanities, and to concentrate their efforts toward writing for other professional historians of science.[5] And it is true, of course, that the professional historian of science will have professional motivations that can sustain him through books like Dijksterhuis's and even more unreadable ones, without need of the niceties of the explanatory style.

My reference to Kuhn, Cohen, and Agassi is not, of course, meant to constitute documentation for my case that much of the historiography of science is dull, unreadable and irrelevant. To provide such evidence is, from the nature of the case, probably impossible. This is especially true since the terms *dull* and *irrelevant* are ill-chosen. No doubt the phrase "logico-stylistic divergence from Koyré-Guerlac" would be more fortunate. There are, however, two general reasons which may be adduced in support of the idea that it is intrinsically more difficult to make histories of science readable. The first is that the rate of scientific changes is great, at least compared with the socio-political changes and developments dealt with in general history.

Hence, the difference between the science of even Galileo and Lavoisier, not to mention Descartes or Priestley, and the science of contemporary scientists is much greater than the difference between the behavior of comparably spaced statesmen, for example.

The second reason stems from the growth of knowledge. Besides undergoing great changes, science experiences progress in a much more straightforward and uncontroversial sense than society does; hence past science loses much of its relevance and interest.

Agassi seems more inclined to lament this unreadability than to explain it. In fact he tries to account for it[6] by saying that because of their uncritical acceptance of philosophy of science principles which they are constantly using in their explanations, historians of science fail to explicitly mention and discuss those principles. I have criticized Agassi's claim on the grounds that historians of science do not need any "philosophy of science principles" in their explanations and on the grounds that to neglect to mention and discuss admittedly "problematic and controversial" principles would lead (other things being equal, e.g., assuming the explanations are satisfactory on other grounds) to explanations that engage our attention, rather than discourage it. One may here add an additional refutation of Agassi's claim: Guerlac's work is comparable to Koyré's in its readability, despite the difference of philosophical sophistication.

On the other hand, if we accept Scriven's analytic principle of explanation and if my interpretation of that principle is in fact right, the tediousness and unreadability of historiography of science must be due to the at-least relative absence of historical explanations. Because my interpretation of explanation provides me with a logical foundation for psychological interest and readability, I am inclined not to lament the unreadability but to regard it as indicative of the *nature* of historiography of science.

This is not to say that special efforts should not be made to practice the explanatory logical style. Rather, it is to take the realistic view that increased readability cannot be attained unless special efforts to use this explanatory style are made.

To be exact, if the nature of explanation is what Scriven claims, historical explanations cannot be totally absent from much of the historiography of science. In fact what is and what is not understood, as well as what reduces one to the other are relative, though intersubjective and objective, matters; it depends on the context. The possibility hence arises that the most unreadable work or part of it may be an explanation, that is to say, it may perform the explanatory function, which is to provide understanding when it is lacking.

A personal experience will illustrate how an unreadable work can become explanatory. A recent account of the development of astronomy can be found in Peter Doig's *Concise History of Astronomy*,[7] a work surveying in 309 pages the whole development of the science from its prehistoric or semi-historic beginnings to the twentieth century. In spite of its conciseness the book is much more easily consulted than read; like Dijksterhuis's work, it lacks an explanatory style. Doig's accounts of the various astronomical discoveries do not really have the character of explanations. In particular this is true of his account of the discovery of the planet Uranus (pp. 115–17); yet this account became an explanation for me.

Some time ago I read Kuhn's *Structure of Scientific Revolutions*. In the chapter entitled "Revolutions and Changes of World View" Kuhn discusses "the minor paradigm change forced by Herschel"[8] as an example of an astronomer's "shift of vision"[9] or one of those "transformations in the scientist's world the historian who believes in such changes can discover,"[10] or one of the "many examples of paradigm-induced changes in scientific perception."[11] Reflecting on Kuhn's interpretation of Herschel's discovery led me to ask: Why should Herschel be credited with the discovery of the planet Uranus when it was the Russian mathematician Anders Johann Lexell (1740–1784) who first proved that the comet which Herschel believed to have discovered was actually a new planet exterior to Saturn? And how can Herschel have thereby caused a paradigm change since the discovery seems to be a classic case of paradigm articulation, specifically the application of Newtonian celestial mechanics to a new object? The understanding which I lacked was provided by Doig's account,[12] which I consulted because of Kuhn's reference to it in his brief statement of the discovery.

I learned there how original and novel Herschel's instruments and methods of observation were: that his telescopes were much more powerful than those constructed up to then and that he systematically observed every portion of the sky, something which had never been done before. Although Uranus's motion could have never been detected without Herschel's instruments and methods, once the motion had been detected, only computation was needed to determine the nature of the planet's orbit. At the time that Lexell made his calculations, which would have soon been made by someone else if not by him, other mathematicians, including Laplace, had already shown that the motion constituted no ordinary cometary orbit. Of Herschel's technical innovations the same cannot be said. I also learned that a great sensation was caused by Herschel's discovery, which in fact is the first discovery of a planet made in historic times. The facts contained in Doig's account made comprehensible what I initially did not understand.

My conclusion is that Doig's account was the explanation for me at the time. But is this not a purely subjective and biographical fact, having nothing to do with the logic and methodology of history-of-science explanation? Not at all. For one thing, I hope that the reader himself has reached the same conclusion, that the account was the explanation for me at the time. And if he has reached it, I think he does so on the grounds that, in the context discussed above, it is an objective logical fact that there was a problem in my mind, and that Doig's account, as sketched, solves that problem. Unless this were so, Doig's account would not have been an explanation, even for me at the time.

I think all we can say about Doig's account in this context is that it has the potentiality of being an explanation, and thus of possessing the interest, relevance, and readability of an explanatory account. But then it is obvious that any work, no matter how unreadable, has that potentiality. Are we then to conclude that explanation is then simply a logical feature of the *mind* of the reader or historian, with the consequence that even Guerlac's and Koyré's books are not in themselves explanatory? This conclusion need not be drawn since it is possible to draw a distinction between potentially and actually explanatory works. The distinction can be made on the basis of the presence or absence of a statement of the problem which explicitly defines the lack of understanding and thus forces the reader to be puzzled, even if he is not already so, or recreates the puzzle even if he once solved it. Of course, the reduction of the not-understood to the understood should also be explicitly present, unless that reduction is obvious. To restate, those works are actually explanatory where we find all three of the elements of an explanation: (1) something which is not understood, (2) something which is understood, and (3) a reduction of the first to the second.

It is possible, then, to structure history-of-science works in such a way that they contain actual explanations, that is, in such a way that the accounts contained therein are explanations for virtually everyone who can read them. Those works will then not be just potentially interesting, that is, interesting when and to the extent that they are consulted for an answer to a problem one may have formulated himself or encountered elsewhere; they will be intrinsically interesting. Finally, I do not want to deny that there are degrees of acuteness in not comprehending and, consequently, of the fascination of explanations. But the degrees can be appreciated best in terms of the ideal types between which they are located.

In order to give explanatory structure to works of history of science, the special efforts to be made are then to practice a logical style which defines what is not understood and then makes it understood, uncovering in the process other things which are not understood and in time rendering these comprehensible.

We are now ready to see how scholarship can bridge, through explanations, the gap between chronicle and history. In Croce's view[13] what characterizes history is the mental attitude of the historian. What distinguishes history from chronicle, for example, is not anything present or absent in the work under consideration, but rather the presence or absence of the historical attitude in a man's mind and practice. To scholarship Croce assigns the necessary function of collecting, rearranging, and purifying documents and evidence and compiling narrations,[14] whose artifacts are kept in places called libraries.

Since history, for Croce, is what provides understanding when a lack of it is mentally or practically felt, the most erudite, scholarly, and literarily polished work is still chronicle if and when it does not or cannot fulfill that function. If I am right in claiming, as I have argued above, that there is a way of compiling a narration which first makes a lack of understanding mentally felt to the reader and then provides him with the understanding, then there is a type of scholarship which can claim to be history or at least approximate it.

Satisfactory Scholarly Explanation

If, then, it is possible to compile histories, that is to say, if to give a scholarly explanation is not a contradiction in terms, how can one do it satisfactorily? A comparative analysis of the criticized explanations in relation to each other should reveal this, just as the analysis of those explanations as a whole in relation to works of a different kind revealed their truly historical character. I mentioned above that the three explanations criticized were not unsatisfactory in the same way. The main problem with Guerlac's explanation is that Lavoisier simply was not in possession of most of the knowledge and ideas that Guerlac attributes to him. Following Michael Scriven's semi-technical terminology, after having already used his concepts implicitly,[15] I shall call this aspect of deficiency *inaccuracy*. The main difficulty with Koyré's explanation is that the connections between the various elements of Descartes's and Galileo's mental attitudes and their failures, errors and success simply do not exist. Again using Scriven's concepts and terminology, we shall say that Koyré's explanation is "inadequate." Finally, the problem in explaining the rise of modern science is that it is not clear what types of explanations are appropriate and what inappropriate. It is not clear that external factors are appropriate to explain the kind of historical development that the rise of modern science exemplifies, and it is not clear that internal factors are appropriate to explain all the aspects of that historical development.

Thus two respects in which there can be no question of Guerlac's explanation being satisfactory are its adequacy and its appropriateness. Lavoisier did not in fact possess the knowledge and ideas Guerlac attributes to him in the summer of 1772, but if he had, that would explain his performance and interpretation of the experiments mentioned in the sealed note. The connection between his knowledge and ideas and the experiments is adequate to account for them. It is comprehensible how and why Lavoisier would have made and interpreted those experiments as he did if Guerlac's attributions were accurate. And, needless to say, the factors mentioned by Guerlac—Lavoisier's beliefs, knowledge and reasons—are appropriate. For the sake of giving some content to this appropriateness we may mention some inappropriate factors, Lavoisier's childhood toilet training and Lavoisier's economic condition.

As regards appropriateness, no real objection can be raised against Koyré's explanation either. It is also largely accurate in the various features that it attributes to Descartes and Galileo. Renunciation of causal explanation does characterize Galileo's thought, but unfortunately at a time (in the *Discourse*) when it makes incomprehensible how he could have succeeded. Similarly, although excessive geometricism constantly characterizes Descartes's attitude, and pure mathematicism his attitude at the time before 1630 when he studied Beeckman's problem, they do not in the least make Descartes's error understandable. First, Descartes never formulated, like Galileo, the geometricist principle of acceleration according to which the acceleration would be proportional to the distance traversed. Second, the problem Beeckman had given Descartes was mathematical; whatever physical misunderstandings Descartes had, they are demonstrably irrelevant to the error he did commit. Finally, it is probably accurate to say that Galileo accepted a physico-mathematical theory of scientific method; but the connection between such a theory and Galileo's practical scientific achievements remains a mystery.

In explaining the rise of modern science, the historian seldom denies —or logically can deny—the presence of various external factors. We may thus say that, in general, it is not the accuracy of the various explanations that is unquestionable. However, we cannot say here, as was the case with Guerlac's and Koyré's explanations, that the explanations are satisfactory in a second sense. Actually, it is true that when the relevance of the external factors is questioned, what one is often questioning is the appropriateness of that type of factor. But sometimes the question of appropriateness is and ought to be debated on grounds other than purely typological ones, namely by consideration of adequacy. For what is often the question is whether a comprehensible connection exists between external and internal factors.

We may conclude by saying that, first, Guerlac's explanation, though inaccurate, is adequate and appropriate; second, Koyré's explanation though inadequate, is appropriate and largely accurate; and third, explanations of the rise of modern science are usually accurate but questioned and questionable in appropriateness and adequacy.

It should not be forgotten that the above-mentioned properties of explanation are not absolute, but as dependent upon context as explanation itself. Thus if unanswerable objections were made against the appropriateness of Koyré's explanation, it would become inappropriate. But, of course, the context works both ways. In fact, I have argued that Guerlac's explanation is adequate and Koyré's appropriate even though they are not explicitly proved to be so; even though no justification is given for the adequacy of the first and for the appropriateness of the second, they indeed have that property. Justifications are simply unnecessary in the context.

Thus it should be obvious that my criticism of Guerlac's and Koyré's explanations neither consists of nor reduces to the allegation that those explanations have not been fully proved. It would be a misunderstanding of the deficiencies in those explanations to think that all my criticism shows is that the evidence given is not sufficient for their acceptance. To repeat, I do not think that there is anything like a complete, that is, a final or ultimately complete, justification or proof of any assertion, explanation or not. I claim that the evidence available in the context points in a different direction from that taken by Guerlac and Koyré. *If* the evidence *had* been right it would have been sufficient for accepting their explanations.

chapter **15**

History-of-Science Explanation:
Its Microstructure

THE PREVIOUS CHAPTER distinguished history-of-science explanation, such as found in Guerlac's and Koyré's works, from works
of scholarship which are not explanations, and marked three ways in
which an explanation can be satisfactory and unsatisfactory. Those
distinctions are of general importance as well as essential for fully
appreciating and evaluating Guerlac's and Koyré's works. However,
they provide little insight into the fine structure of history-of-science
explanation, which is important both for its own sake and for the purpose of determining the similarity between the explanations criticized
above and other historiographical work. I shall now concentrate on the
microstructure of history-of-science explanation.

Following Scriven's analysis[1] in general, though not in detail, the first thing to be done is to distinguish the statement of the explanation or the *explanatory claim,* as I shall call it, from the justification or evidence or arguments in support of the explanation, or *grounds* for the explanation. The general distinction between an explanatory claim and its grounds is relative, being dependent on the contextual nature of explanation itself. To express the distinction in general terms, the explanatory claim is the set of those assertions which if true make comprehensible the thing which is not understood. The grounds are those reasons, arguments, or evidence needed in the context to justify the explanatory claim.

From this distinction and from the three-fold aspect of satisfactoriness, it follows immediately that there may be three kinds of grounds: accuracy-justifying, adequacy-justifying, and appropriateness-justifying. I have said that the grounds may be present—if one is putting forth a justification of an explanatory claim—and they must be produced if and when the explanatory claim is questioned.

Thus the series of quotations cited from Guerlac's, Koyré's, and Basalla's works, respectively were meant to be, or at least to approximate, the various explanatory claims. They were meant to and, by and large, do exclude the grounds. Respecting Guerlac's and Koyré's explanations, I stated and discussed the grounds in my criticism, since those two works are indeed justifications of explanatory claims. The grounds of the explanations of the rise of modern science, on the other hand, I did not discuss since such grounds are usually accuracy-justifying, whereas the main problem concerns the appropriateness and adequacy of those explanations.

Explanatory claims, too, have their own structure, which includes the thing to be explained (the *explicandum*), and the series of statements that does the explaining (the *explicans*).[2] One of the things to be explained is the rise of modern science. What is to be explained by Guerlac is Lavoisier's performance and interpretation of the experiments mentioned in the sealed note. What is to be explained by Koyré is that in the discovery of the laws of falling bodies Galileo succeeded whereas Descartes and Galileo himself previously failed. Guerlac's main explicandum consists of two parts: Lavoisier's performance and his interpretation of the litharge experiment, and his performance and interpretation of the sulphur and phosphorous experiments. Koyré's explicandum consists of three parts: Galileo's early error, Descartes's failure, and Galileo's eventual success. Moreover, the things that explain these three explicanda are distinct, though of course related.

Taking as basic and without defining the notion of historical development we may begin by saying that all three explicanda are historical
developments. Yet the rise of modern science is very different from the
other two. First of all it is a development directly involving a large
number of individuals, and we may thus call it a social development.
Lavoisier's discovery on the other hand directly involves only himself,
and the discovery of the law of falling bodies directly only Galileo, Descartes, and perhaps Beeckman. Of course the two discoveries involve
indirectly large numbers of individuals; for example they will affect
many contemporaries and successors and they are in many ways dependent on many contemporary and preceding individuals. The difference still remains, however, because the rise of modern science also
involves indirectly many more people besides the participants. In other
words, practically all historical developments involve in some way or
other large numbers of persons, at least after the time of the development; but social developments are the ones that directly and primarily
involve large numbers of persons; individual developments one or very
few.

There is another more important difference between the rise of modern science and the two discoveries: the former is unique in an important sense in which the latter two are not. To begin with, it should
be noticed that I avoided referring to the rise of modern science as
"the scientific revolution," let alone as "the scientific revolution of the
seventeenth century." It is not that I have thereby wanted to deny or
minimize the revolutionary character of that development. Rather, I
have wanted to deny the implication that it was one of the many revolutionary developments in the evolution of science, perhaps the single
most important such development, but still one of a kind. I believe that
to begin with such descriptions of the rise of modern science prevents
one from appreciating the special difficulties for the explanation of that
development. Moreover, that description does not do justice to the
special regard in which it has been held by scientists, philosophers,
historians of science, and general historians alike. The theorists of scientific revolutions *begin* with it and argue that many other history-of-
science developments have the same structure; medievalists-traditional-
ists-continuists begin with it and try to show that it originates in some
sense or other from medieval science; scientists still find their heroes
in the leading figures of that period, in spite of the twentieth-century
upheaval in physical theory; philosophers, be they empiricists, posi-
tivists, or Achimedeans, all treat it as the test case of their epistomolo-

gies. All we can say, then, about the nature of this explicandum is that it is in a class by itself.

It contrasts sharply with the other two explicanda; they both belong to the class of discoveries, Lavoisier's discovery of the explanation of the weight-augmentation effect and Galileo's discovery of the law of falling bodies. Having pointed out this similarity and indicated that the characteristic shared by them is also shared by many other individual developments, so that we can expect the explanation of them to be relevant to the explanation of the similar developments, I now note the differences between the two explicanda, which may be characterized by saying that they represent different *aspects* of discovery to be explained. Guerlac's explicandum is a *cognitive* aspect, Koyré's is an *evaluative* aspect. Guerlac is trying to explain the origin of the intellectual content, or at least of an aspect of the intellectual content, of Lavoisier's discovery. Koyré is trying to find the sources of the success element in Galileo's discovery, of the failure/error elements in Descartes's investigations, and of Galileo's earlier error. *Error, failure, success:* all are terms connoting the good or the bad. That the cognitive aspect of discoveries is a common subject for historians of science to investigate and explain needs no special emphasis. That historians of science also deal frequently with the explanation of the evaluative aspects of discoveries has seldom, if ever, been explicitly noticed. To refer to the nature of the explicanda, Guerlac's explanation will be called an *explanation of a scientific cognition,* Koyré's an *explanation of a scientific success and of two scientific failures.*

The explicans of Guerlac's main explicandum consists of the intellectual genesis of Lavoisier's litharge experiment and interpretation. The primary elements of this genesis are Lavoisier's reasons for his experiment and interpretation: that air plays a chemical role in calcination, that metals increase in weight when calcined, that the phlogistic theory cannot account for this weight augmentation, that calxes effervesce when they are reduced, and that metals resist calcination in closed vessels. To refer to the nature of the explicans, I shall call this part of Guerlac's explanation a *reason-explanation of a cognition.*

Besides this intellectual genesis of Lavoisier's experiment and interpretation, Guerlac's explicans includes an account of the origin of Lavoisier's reasons. For, in the context of Guerlac's explanation, it is not immediately comprehensible how Lavoisier could have come to the beliefs and knowledge which constituted reasons for the experiment and interpretation. The genesis of these reasons is, in Guerlac's account, primarily nonintellectual, but it is no less appropriate. In fact, once a cognition has been traced to its intellectual roots, it is not just proper, but often, as in this context, imperative, that one should search for its possible nonintellectual sources. At least this search is necessary if one wants to give a complete explanation. In fact, I argued in my criticism of Guerlac that his explanation was incomplete to the extent that he did not include an account of the genesis of Lavoisier's belief that the phlogiston theory was incompatible with, and thereby refuted by, the weight-augmentation effect. I also argued that this incompleteness of Guerlac's explanation was the result of Guerlac's own mistaken belief that the augmentation effect is incompatible with the phlogiston theory.

Just as Guerlac's explanation is incomplete in the above respect, it is complete as regards the origin of Lavoisier's other reasons. And that completeness may be better appreciated if contrasted to the incompleteness of Koyré's explanation regarding the origin of Descartes's and Galileo's methods or mental attitudes. But first I shall have to make a qualification about the nonintellectuality of Guerlac's sources for Lavoisier's motivating reasons. Guerlac sees Lavoisier's conviction that air plays a chemical role in calcination as originating partially in Lavoisier's more general idea that air plays an important role in chemical processes. Guerlac attributes both an intellectual and what might be called a *practical* origin to this latter idea, that is, an origin in Lavoisier's experimental investigations of the early summer of 1772. Finally, he attributes Lavoisier's suspicion to the influence of "Hales —indirectly through Boerhaave, Rouelle, Macquer, and others, if not directly through a reading of the [book by Hales] *Vegetable Staticks*";[3] whereas Lavoisier's experimental investigations of the phenomenon of effervescence are seen as the result of the appearance in France of Priestley's *Directions for Impregnating Water with Fixed Air*. Analogously, according to Guerlac, Lavoisier's knowledge of the augmentation effect and of the resistance of metals to calcination in closed vessels derives from reading Guyton de Morveau's *Academic Digressions*. As for the effervescence of calxes during reduction, Lavoisier probably found it described in the chemical literature of the day. The part of Guerlac's account in which he traces the origin of Lavoisier's reasons may be called a "socio-psychological explanation of cognition." It should be noted that in my criticism of Guerlac's explanation I accepted his socio-psychological account and limited myself to criticizing his reason-explanation.

The most important characteristic of the explicantia of Koyré's three explicanda is that they consist primarily of various general features of Descartes's and Galileo's thought and behavior, which may be termed *methods*. They are: geometricism (excessive geometrization) and renunciation of causal explanation to explain Galileo's early error, geometricism and pure mathematicism to account for Descartes's failure, and Galileo's physico-mathematical mental attitude to explain his success. To refer to the nature of the explicantia Koyré's explanations may be called *method-explanations*. More specifically, Koyré's account of Galileo's error and his account of Descartes's failure are two *method-explanations of scientific failures*, and the account of Galileo's success is a *method-explanation of a scientific success*.

I have already mentioned that Koyré, unlike Guerlac, does not, by and large, try to explain how and why Galileo and Descartes came to use, accept, and practice those methods. To give content to my qualification "by and large" and substance to the implicit charge of incompleteness, it should be pointed out that Koyré actually does attempt to make comprehensible the alleged fact that Galileo's early thought had a geometricist character. He does so in terms of the natural tendency to visualize in space rather than to think in time, the intrinsically greater intelligibility of space as contrasted to time, and the fact that, before Descartes's analytic geometry, any mathematization had to be geometrization. But we find no explanation of Galileo's renunciation of causal explanation, of Descartes's pure mathematicism and geometricism, and of Galileo's physico mathematicism.

It should be noted that for two reasons we cannot account for Descartes's geometricism in the same way as Galileo's: analytic geometry was available to Descartes, and his geometricism took a different form from Galileo's. Thus we cannot claim, on the grounds that explicit repetition of the explanation is obviously unnecessary, that Koyré really accounts for Descartes's geometricism when he explains Galileo's. At any rate, to make the claim would only increase the completeness of Koyré's explanation at the cost of decreasing its *adequacy* since, as I have mentioned, what explains Galileo's geometricism cannot explain Descartes's.

Another way of increasing the completeness of Koyré's explanation, but at the expense of other qualities, would be to argue that in the given context it is not necessary to trace the origin of the various methods followed by Descartes and Galileo. This could be done by arguing that it is well known that the methods enumerated characterized

Descartes and Galileo and, thus, in the given context, it can be taken as understood that Descartes and Galileo used them, so that any reduction to them of whatever is not understood is a complete explanation. It is questionable whether at the time of the second volume of *Etudes Galiléennes* (1939) it was well known that these methods characterized Galileo and Descartes. But at any rate the effect of the above completeness-justifying argument on Koyré's explanation would be to decrease its merit. If Koyré's work, like Guerlac's, is the justification of a complex explanatory claim, and if the only justification attempted is to show that the various successes or failures are due to methods the presence of which is taken as unproblematic, then Koyré's explanation is a failure since it fails to connect the success and failures to those methods. In other words, one can claim that Koyré's explanation is complete in the above sense only at the cost of claiming that Koyré's work was not the first to draw attention to the above mentioned features in the thought of Descartes and Galileo.

To conclude this phenomenological analysis of the microstructure of explanation one may think of explanatory claims as propositions of the form "p because q" or combinations thereof. In fact, almost always, with more or less artificiality, one can put an explanatory claim into that canonical form. That form will seem less strained if one admits that both p and q may stand for compound sentences of any order of complexity. To emphasize this we might say that the really general form of explanatory claims is "$p_1, p_2, p_3, \ldots$, and p_n because $q_1, q_2, q_3, \ldots$, and q_k." A common and intermediate case is a claim of the form "p because $q_1, q_2, \ldots$ and q_k," where p refers to *one* (easily identifiable) thing to be explained and each q_i to one (easily identifiable) explanatory factor. The p_i's refer to or are the explicandum, the q_i's the explicans.

For example, Guerlac's central explanatory claim can be stated in this way: Lavoisier performed the litharge experiment because he wanted to determine whether it was indeed true that the air fixed in metals during calcination and released upon reduction accounts for the greater weight of a calx; and Lavoisier had conceived of the latter hypothetical possibility because he was convinced that air plays a chemical role, and he knew that metals increase in weight when calcined, that they resist calcination in closed vessels, and that calxes effervesce during reduction. The central explanatory claim in Koyré's explanation of Galileo's error is the following: Galileo postulated the wrong principle of acceleration because his thinking was characterized by excessive geometrization and he had given up the attempt to causally explain the phenomenon of fall. The central explanatory claim in the explanation of Descartes's failure has the following form: Descartes failed to derive the correct time-distance formula because he approached the problem with the attitude of pure mathematicism and his thinking was beset by a tendency toward excessive geometrization. Finally, according to Koyré, Galileo succeeded because he had a physico-mathematical mental attitude.

I shall now show (1) the microstructural similarity between Koyré's work and other accounts to be found in the contemporary or recent literature of the history of science and (2) the existence in the latter of method-explanations of successes and failures. I shall do this by taking various passages from history books and extracting from them the various explanatory claims in canonical form. To do the same for Guerlac's work is, I believe, unnecessary because the existence of what I have called reason-explanations of cognitions is well known.

The first four items are taken from A. R. Hall's *Scientific Revolution.* First (pp. 339–40) Hall states method-explanations of Lavoisier's general success, of one of Lavoisier's failures, and of the general failure of Boyle's chemistry. They can be expressed as follows: Lavoisier succeeded in his chemical investigations in general because he relied more heavily on quantitative procedures, he coordinated theoretically the work of the great experimenters which was well known and accepted by many, and he treated chemical phenomena independently from the physics of the time. In fact, Boyle's chemistry was by and large a failure "because he tried to explain chemical phenomena rigidly in terms of a physical theory," and Lavoisier himself failed in his theory of caloric because he tried "to bind physics and chemistry prematurely in one." [4]

Second, there is a passage (p. 176) which is Hall's closest approximation of a method-explanation of the rise of modern science. It can be paraphrased as follows: the scientific revolution flourished because it was discovered that explanation and description have no really distinct significance in science. I believe this means that explanations are not something different from descriptions, but rather they are descriptions which perform a certain special function—the function of providing understanding, one might say following Scriven's theory. And descriptions are not something inferior to explanations, but rather they are what explanations become when the explanatory function and context are not present. (Note also the analogy with Croce's view of the relation between history and chronicle.)

Third, Hall claims (p. 183) that, to the extent that Descartes succeeded, he succeeded not because his method was a good method, but because he was a genius.

Fourth, Hall claims (p. 184) that some progress in seventeenth century science was made because hypotheses were framed freely "with rigorous attention to the findings of experiment and observation." This claim, besides being an example of one of Hall's method-explanations of success, also illustrates a frequent characteristic of his statements which may be called *qualification to inconsistency*.

The next three claims are found in the second volume of A. C. Crombie's *Medieval and Early Modern Science*. Crombie first gives (p. 162) the following method-explanation of Galileo's success: Galileo achieved his success because he eliminated the physical-causal elements from the problem of motion and tried "to solve each individual problem separately, to discover empirically what laws were in fact exhibited by the natural world, before facing the task of reassembling them as a whole." [5]

Second, Crombie claims (p. 161) that "Descartes completely failed to grasp the essential concept of the conservation of momentum" because his thinking was geometricist, that is, because for him "the real world was simply geometry realized; movement he conceived of simply as a geometrical translation, with time as a geometrical dimension like space."

Third, Crombie gives (p. 121) the following method-explanation of the rise of modern science: "The scientific revolution of the 16th and 17th centuries came about" because men began "asking questions within the range of an experimental answer . . . limiting their inquiries to physical rather than metaphysical problems, concentrating their attention on accurate observation of the kinds of things that are in the natural world and the correlation of the behavior of one with another rather than on their intrinsic natures, on proximate causes rather than substantial forms, and in particular those aspects of the physical world which could be expressed in terms of mathematics."

The eighth claim is drawn from of C. C. Gillispie's *The Edge of Objectivity* (pp. 93–94): Descartes "spun so crude a physics" because he thought that nature herself, not just the laws of nature, is simple, his thinking was excessively mathematical, and Cartesian science was too ambitious in that it tried both to describe and to explain phenomena with the same causal-mechanistic account.

I am here primarily concerned with showing the frequency of explanatory claims of the same *kind,* i.e., with the same fine structure, as Koyré's claims in Volume II of the *Etudes Galiléennes.* However, one cannot help noticing how substantively similar those claims are; that is, the content of those claims, the methods by reference to which successes and failures are explained, goes very little beyond Koyré's work. But though Koyré's followers follow him in both form and content in this respect, other historians make claims with the same microstructure.

For example, J. F. Scott, in *The Scientific Work of René Descartes* claims (p. 162) that Descartes failed in his formulation of the laws of motion because his concept of motion had no mathematical appeal in that it did not, for example, quantify the exchange of motion between two bodies. He also claims (pp. 163–64) that Descartes's science was a failure because he tried to found his system of mechanics upon divine immutability, he had "a readiness to accept as self-evident propositions the truth of which only experiment can decide," and he failed to recognize the decisive importance and correct function of experiments, which he did not reject but whose function was for him that of verifying conclusions which he had reached a priori, thus making a minimum of experiments suffice for a maximum of inference.

In Louis T. More's *Isaac Newton: A Biography,* we find (p. 286) the following method-explanation of Newton's success: "because he limited his scientific work to the formulation of laws, and restrained his imagination from making hypotheses as to how nature operates, his great discoveries and laws are classic and permanent."

Finally, in his *Metaphysical Foundations of Modern Physical Science,* E. A. Burtt states (p. 208) that it is unfortunate that Newton himself does not give us method-explanations of his own scientific successes.

There should be little doubt now that the fine structure found in Koyré's explanation is duplicated in many other historical works. The prevalence of that structure is even more universal than the above evidence indicates. In fact, the accounts of the scientific method of individual scientists which are found in the history-of-science literature are usually the explicantia of method-explanations of successes or failures; Hall's account of Galileo's scientific method and Crombie's account of Descartes's, which I shall discuss later, are examples of such explicantia. To give content to this assertion I note that it is both possible and sometimes the case for those accounts to have a different function. Sometimes accounts of a great scientist's scientific method have the function of confirming a methodological principle that the writer is propounding. And sometimes accounts of an individual's scientific method are used as evidence for evaluating the scientist. For example, one might want to give an account of Copernicus's scientific method to show that in no important sense is he a modern scientist. Or one might want to give an account of Kepler's method to discredit him as a scientist; or again, one might want to give an account of Stahl's scientific method to show the correctness of Priestley's opinion of him, namely, that he was one of the greatest scientists who ever lived.

chapter **16**

Discovery *vs.* Justification,
and Theory *vs.* Practice in Science

HISTORIANS OF SCIENCE have failed to be critical or clear about two distinctions: the first, between the context of discovery and the context of justification; the second, between the context of practice and the context of theory.

The distinction between the contexts of discovery and justification is, in the case of the cognitive aspects of discoveries, the distinction between the individual's reasons for arriving at the cognition and the reasons whereby he justifies it. For example, if I am right about Lavoisier's discovery, his main reason for making the litharge experiment was to elaborate his (mistaken) theory of elements by measuring the amount of air which on the basis of that theory he believed would be absorbed by the calx. On the other hand, after he has made the experiment and learned from it he justified it with the reason cited in the sealed note: he was testing his antiphlogistic explanation of the augmentation effect. A clearer example is Galileo's discovery of the time-distance law. It is obvious that the reasons whereby he arrived at the law of which he was in possession ever since 1604, as his letter to P. Sarpi shows, must have been very different from the ones whereby he justified that law, either tentatively and erroneously in that same letter, or eventually in the *Discourse*.

In the case of the evaluative aspects of a discovery, the distinction is one between the way or method by which one arrives at the cognition which is being evaluated and the way or method whereby one justifies the evaluated cognition. We need not elaborate here the connection between reasons and methods, but it may be pointed out that one of their differences is in the greater generality of methods. A method might consist in giving reasons of a certain kind; it could hardly consist in giving the specific reasons once given by someone on a given occasion; however, one very general method is that of giving reasons, pure and simple, and it may be called the *rational method*.

To be clear about the general distinction between contexts of discovery and of justification is not necessarily to be able to give a general characterization of the distinction which eliminates all problems about borderline cases. I have just given such imperfect characterizations which will suffice for our present problem. However, the best way to make the distinction is by using the notion of explanation itself. The reasons of the context of discovery, which I shall term the *logic of discovery*, are those which are present in a satisfactory reason-explanation of a cognition. The reasons of the context of justification, which may be termed the *logic of justification*, are not directly relevant to reason-explanations of cognitions. The logic of justification may be indirectly relevant by allowing us to find or understand better the logic of discovery; but the logic of justification cannot, qua reasons of the context of justification, constitute the explanation of cognitions. It is possible and not infrequently the case that the logic of discovery and of justification should coincide, and when this happens the logic of justification may constitute the explanation of a cognition. Similarly, the following connection holds: the methods of the context of discovery, which may be termed the methodology of discovery, are those which are present in a satisfactory method-explanation of success; the methodology of justification, on the other hand, is not, as such, relevant to that kind of explanation.

The reality of the distinction is, I believe, a more or less direct consequence of the growth of knowledge. In fact, the situations to which the distinction is relevant are situations in which learning is taking place: individuals are learning either by thinking or experimenting or observing. The learning that the individual experiences cannot fail to affect his view of cognitive experiences which he had before the learning occurred. Thus, in learning situations, the reasons which an individual has for holding a certain belief are time dependent. His present reasons are his justification of his belief, they are the answer to the *logical 'why' question,* "Why does he hold that belief." But besides this question, one can ask the *causal-temporal 'why' question,* "How did he come to hold that belief." This general divergence of logic of discovery from the logic of justification results in a divergence of the respective methodologies, since methods to a large extent derive from reasons. The situation is further complicated by the logical and methodological theory which the individual may hold. It is possible, for example, to imagine someone who may accept a logical theory of justification according to which the logic of justification should be made identical to the logic of discovery. Such a person would answer the logical 'why' question, "Why do you hold such and such a belief," by giving a statement of how he came temporally-causally to hold that belief. If Kepler had held a logical theory conforming to his practice he might have come close to being such a person. This brings us to the distinction between the contexts of practice and theory.

The distinction between practice and theory is that between what an individual does and what he says, between his deeds and his words. The context of theory is the context of reflection, the context of practice is that of involvement. This distinction cuts across the previous one so that there can be theories of the logic or methodology of discovery, and logical or methodological practices of discovery; and there can also be theories of the logic or methodology of justification, and logical or methodological practices of justification. The theory-practice distinction within the context of discovery is, however, the more important one because it is here that the divergence is most notable. It is commonly known, in fact, that what scientists say about their reasons for and methods of discovery is usually very different from the reasons they had and the manner in which they made the discovery. What often happens is that what one says about one's logic and methodology of discovery coincides with one's practice of justification.

In general, it is very difficult to distinguish between a scientist's words and his deeds. In some trivial sense all that is available to the historian about scientists who are already dead is their words. The distinction does not vanish, however, because within the set of a scientist's writings there are contexts where he was involved in what he was doing, and there are others where he is reflecting on what he has done or is doing. For example, Lavoisier says in the sealed note that his discovery that both sulphur and phosphorous increase in weight when burned because they absorb air "made him think" that the increase in weight of metals during calcination "had the same cause." The context there is obviously one of theory; Lavoisier is reflecting on what he did and reporting it to us; the word character of Lavoisier's claim remains such even if what he says should happen to be true, which it is not, as both Guerlac and I would agree. The present distinction although not absolute is real nevertheless.

Guerlac's Explanation

Before elaborating these distinctions any further or defending them against possible objections, I want to show how the failure to be critical or clear about them leads Guerlac and Koyré to their unsatisfactory explanations. The first thing to be noticed is that one great merit of Guerlac's explanation and one respect in which it is superior to Koyré's is that Guerlac is aware of the distinction between the contexts of theory and of practice. In fact, Guerlac's whole explanation is based on his skepticism about what Lavoisier tells us was the order of the phosphorous, sulphur, and litharge experiments and his reason for conceiving the antiphlogistic explanation. Unfortunately, Guerlac's use of the distinction between theory and practice is opportunistic. I do not object to the apparent inconsistency of Guerlac's attitude, for the inconsistency may indeed be apparent and not real; rather I hold that he should have been critical in his use of the distinction between the context of theory and the context of practice. To be thus critical would have consisted in giving reasons why Lavoisier's words should be trusted when Guerlac was trusting them, just as he does give some reasons for not trusting Lavoisier's words of the sealed note.

What I have just done is to turn one of my earlier methodological criticisms of Guerlac's explanation into a causal explanation of its deficiency. The methodological criticism was different from the substantive and specific objections in that it summarized and abstracted from them thus striking at the root of Guerlac's error or failure. The explanation of the error gives the cause as being the method, the presence of which made the error not just objectionable, but methodologically so. I can now proceed more explicitly to use my third methodological criticism, which referred to Guerlac's frequent confusion between what I called historical and logical issues. That was simply a less technical and presumably less clear terminology for what I can now call a confusion between the context of discovery and the context of justification.

My second methodological criticism was that Guerlac's explanation was "overcoherent." That is to say, it was fundamentally misconceived to explain both the performance and the interpretation of the litharge experiment, at one and the same time, by making the experiment a test for the explanation of the augmentation effect. In the light of a critical distinction between contexts of discovery and justification, the confirmation of the antiphlogistic explanation by the litharge experiment mentioned by Lavoisier in the sealed note would have been recognized as reflecting Lavoisier's logic of justification and not that of discovery. That is, the sealed note would have been interpreted as saying not that the litharge experiment was made after tentatively explaining the augmentation effect, but that it can be used to justify that explanation. In fact, public pronouncements are normally in the context of justification since it is justification of one's ideas that is of public interest. And the sealed note, though sealed, was intended for eventual public disclosure. In other words, the logic of justification coincides to a large extent with the logic of public written exposition. Or at least the context of exposition is as good a sign of the context of justification as the presence of expressions like "I remember that . . ." or "what convinced me was . . ." are signs of the context of theory. If Guerlac had been clear and critical about the distinction between contexts of discovery and of justification he would have doubted that Lavoisier's logic of discovery coincided with his logic of justification, and would have been skeptical, or more critical, about evidence supporting the coincidence. He would then not have been so quick to read Turgot's idea into the last section of the August memorandum.

Koyré's Explanation

An equivocation runs through Koyré's explanation. The law of falling bodies is normally and properly to be taken to be the time-distance formula according to which the distances traveled by a body falling freely from rest varies as the square of the time elapsed. But Koyré explicitly states the law (p. 84) as follows: free fall is a uniformly accelerated motion (uniformly with respect to time). For purposes of identification we may refer to this latter statement as the principle of acceleration. The law is a special case of the principle. One problem with Koyré's explanation is that whereas Galileo's error and success take place in the context of the discovery of the principle, Descartes's error is an error in the context of the (missed) discovery of the law. Koyré never discusses the context of Galileo's discovery of the law and he explicitly recognizes the omission (p. 87). What Koyré does discuss is Galileo's methodology of discovery of the principle of acceleration, which coincides with his methodology of justification of the law. If Koyré had attended to these distinctions he would have been more careful about exactly what errors he was considering and he might not have been so rash as to claim the various connections with the various methods. Those connections I showed to be usually nonexistent.

I have said that Koyré discusses Galileo's methodology of discovery of the principle of acceleration. More specifically, that is what Koyré *should* be discussing and what he needs to account for Galileo's early failure and eventual success in the discovery of the principle. But Koyré's unawareness of the distinction between the context of discovery and the context of justification leads him to take the methodology he needs from the context of justification. No wonder, then, that we are unable to find any connection between the methodology Koyré attributes to Galileo and either his error or success.

No doubt Koyré is building his whole case on certain elements of Galileo's methodological practice, relating directly or indirectly to the notion of time and to temporal considerations. The clearest indication of this is the importance that Koyré attaches to Galileo's remarks about the "supreme affinity which exists between motion and time." [1] Moreover, the methods that he discerns in Galileo's earlier investigations and in terms of which he tries to account for his error are geometricism and renunciation of causal explanation. And in the course of Koyré's unsuccessful argument to connect these methods with Galileo's error they both turn out to be manifestations of a disregard of the supreme affinity between motion and time. According to Koyré, in fact, Galileo's early geometricism is simply intemporalism; it is the choice of spatial over temporal acceleration; and renunciation of causal explanation is interpreted, by equating causal with temporal considerations, as failure to engage in temporal considerations. Finally, the methodological practice in terms of which Koyré explains Galileo's success is the alleged constant attention paid by Galileo to the real character of the phenomenon, and for Koyré "real" here means simply temporal. And the methodological theory that in the same discussion (p. 156) Koyré attributes to Galileo needs no separate argument here insofar as Koyré tends to identify it with the above mentioned practice of "constant attention."

The basis of Koyré's whole explanation is then his attitude toward Galileo's remarks about the supreme affinity between time and motion. From these remarks he builds up Galileo's methodology of discovery of the principle of acceleration. However, an examination of the context in which those remarks were made by Galileo shows that it is one of justification. In the passage quoted by Koyré (pp. 136–38), Galileo is trying to make plausible, to justify, the principle of temporal acceleration; he is trying to show that, as the last sentence quoted by Koyré concludes, "from the preceding arguments it appears that we will not deviate from right reason if we admit that the intensity of the velocity [i.e., the instantaneous velocity] increases with the extension of time" (p. 138). And the main preceding argument is that if we pay attention to the affinity between motion and time we will realize that the simplest notion of uniform acceleration is one such that the increments in velocity occur in the simplest possible manner, and nothing is simpler than to have equal increments of velocity occur in equal intervals of time. The strained nature of the argument and the artificial connection with simplicity alone should have made Koyré skeptical about whether Galileo could have come to accept the principle by following that logic and the methodology embedded in that logic. An awareness of the distinction between contexts of discovery and of justification would have been sufficient to prevent Koyré from both improperly grounding his claim about Galileo's method of discovery on the latter's methodology of justification and from disastrously constructing the former out of the latter. "Disastrously" because, by so doing, he is led to claim the existence of certain con-

235

nections between that methodology and Galileo's success and error; and the connections he needs in his explanation are simply lacking.

One difficulty demands a solution, before we proceed further. The Galilean passage mentioned above is found in a fragment, the date of which is probably about 1609. Does not then the justification contained therein acquire an aspect of Galileo's methodology of discovery of the principle when looked at from the point of view of the *Discourse* and of 1638? The reason would be that if as early as 1609 Galileo is trying to justify the principle, that kind of argument must have played some role in his achievement of success in the *Discourse*. In other words, the distinction between context of discovery and context of justification is not absolute, and thus what is justification in 1609 is part of the process of discovery of the finished science of the *Discourse*.

It is indeed true that the present distinction is relative, but to say that is not to say that it is arbitrary. At any rate, even if the above argument was acceptable, it would only show that some role was played by Galileo's attention to time, not that this attention played the central role alleged by Koyré. The argument is not acceptable, however, because as Koyré himself admits (p. 136), the passage found in the 1609 fragment is reproduced in the *Discourse*, and in the *Discourse* the context is one of justification. Hence, the argument to which Koyré attaches so much importance is simply a way of making the principle of acceleration plausible, a way of arguing with respect to which Galileo did not change his mind from 1609 to 1638.

There is indeed an aspect of Galileo's logic of justification from which parts of his methodology of discovery can be retrieved, but it is far removed from Koyré's conception. Galileo discovered the principle of acceleration because of his attempts to derive the time-distance law from a self-evident principle. In 1604 Galileo knew that the time-distance formula was true but did not understand why it was true. His investigations were an attempt to reduce what he did not understand to something that he understood. If all that he had wanted to do was to axiomatize, that is, to derive the time-distance law and the other descriptive formulas from a few general principles, independently of the comprehensibility of these principles, then he would not have felt the need to make plausible the principle of acceleration, still less to make it appear simple. Some aspects of Newton's thought can be interpreted as being little more than axiomatization. But Galileo wanted to find an explanation, in Scriven's sense, of the time-distance law. Galileo's concern for explanation is then part of the methodology he used in his discovery of the principle of temporal acceleration.

Finally, the physico-mathematicism or Archimedeanism that Koyré attributes to Galileo and with which he tries to account for the latter's success look like a methodological theory. For Koyré claims (p. 156, my italics) that Galileo "begins with the *idea* . . . that the laws of nature are mathematical" and "*tells* us to begin with experience" and "*tells* us to be guided by the idea of simplicity." Moreover, some of the evidence Koyré uses is verbal evidence from Galileo, such as his claim in the same passage mentioned above where he tries to make plausible the principle of acceleration. Galileo says that "in the investigation of the definition of uniformly accelerated motion we were guided, by the hand, as it were, by an understanding of the character and habits of nature in all her other works, in which nature is in the habit of employing the simplest and easiest means" (p. 136). It would seem that one can succeed only because of one's methodological practices, because one behaves in certain ways, perhaps in accordance with certain methodological principles. An awareness of the distinction between the contexts of theory and of practice would have led Koyré to at least consider the above mentioned problem. He might have then tried to solve the problem by explicitly constructing an adequacy-satisfying argument to connect the methodological theory with the success. In the interest of seeing what such an argument might be and where it would lead us, let us consider the idea that the laws of nature are mathematical. If we take Koyré's formulation of the law of falling bodies as the principle that free fall is a motion uniformly accelerated with respect to time, then there seems to be nothing particularly mathematical about this principle, as there is, for example, about the time-distance law. Suppose we argue that Galileo's principle of physico-mathematicism tells him the kind of thing to look for and assures him that he need not search any further if, when and after he has found a mathematical relationship. So far the argument may be accepted and would help to explain Galileo's success by contrast to the Aristotelians who accepted a qualitative philosophy of nature. But in the context discussed by Koyré, the argument is worse than useless because the Galilean investigations under consideration, begin, not end, with the finding of the time-distance law. It is not my task, of course, to construct an adequate argument. What I want to conclude, rather, is that if Koyré had been clear and critical about the distinctions between the contexts of theory and of practice with the consequent sensitivity to the difficulty of accounting for successes and failures in terms of a scientist's methodological theory, he might either have been able to construct an adequate argument or if not, he would not have made the present claim, thus avoiding the deficiencies of his explanation. My general conclusion is that Koyré's and Guerlac's explanations are unsatisfactory because they fail to distinguish critically, or at all, between contexts of discovery and of justification, and between contexts of theory and of practice.

chapter 17

"Scientific Method"

HISTORIANS' ACCOUNTS OF "scientific method," although they are not as explicitly explanatory as Koyré's, are microstructurally similar in that they have the fine structure of explicantia of explanatory claims of the form of method-explanations of successes and failures. I shall examine some examples in the light of the two distinctions between the methodology of discovery and the methodology of justification, and between methodological theory and methodological practice. In fact, a conceptual confusion among all four of these things is present in the historians' notion of "scientific method."

Hall's Account of Galileo's Scientific Method

In *The Scientific Revolution: 1500–1800*, A. R. Hall sees four main elements in Galileo's "scientific method": philosophical-mindedness, abstraction, mathematization, and experimental observation (pp. 168–77). What he does not make clear is whether each is an element of Galileo's methodological theory of discovery or methodological theory of justification or methodological practice of discovery or methodological practice of justification.

Regarding his claim that Galileo was philosophically minded, Hall does not have to make clear whether Galileo was being so in his theory or in his practice. For it is obvious that one is referring to his practice since it makes sense to attribute philosophical-mindedness only in the context of scientific practice. But is this practice of discovery or of justification (or both)?

When Hall first states that Galileo was philosophically minded, he is making a claim about Galileo's practice of discovery:

> [Galileo] may properly be described as a philosopher, for his conscious reflection on the obstructions to be overcome in *arriving at* a clear and confident understanding of nature is explicit in a number of passages and implicitly *conditions* the revolution in ideas that he effected. [P. 168, my italics]

But in supporting this assertion, he says that, for Galileo, the relevant philosophical questions

> were not answered in prolonged metaphysical or logical analyses . . . but the answers were given as they became necessary in the progress of his attack on the prevailing ideas of nature. As scientist Galileo's aim might be to detect Aristotle's errors in fact or reason, while as philosopher he demonstrated more fundamentally how these errors had arisen from weaknesses in method that were to be avoided by taking a different course. [P. 168]

That is, Galileo gave answers to philosophical problems when this was necessary to demonstrate that the Aristotelians were fundamentally wrong and that he was right. But this means that Galileo was philosophically minded in the context of justifying his discoveries. It obviously does not prove that he was so minded in his practice of discovery. Hence, either we do not know what Hall is saying, or he has failed to justify his conclusion.

Respecting Galileo's abstraction, Hall does make it clear that it is supposed to be part of Galileo's theory of discovery. That this is so may be seen, first, from the fact that Hall introduces abstraction as the most important element of "Galileo's *remarks on* the procedure to be adopted in *arriving at* these principal generalizations" (p. 169, my italics). Second, Hall claims that Galileo's theory of the method of discovery is of special interest because, as he correctly points out, what we primarily find in Galileo's two greatest treatises is a practice of justification and "it is universally the case that the methods by which a discovery is made and expounded differ, in varying degrees" (p. 168). Presumably it is Galileo's theory of discovery that provides the best access to his practice of discovery. This conclusion, of course, does not follow since it is no less "universally the case" that the theory and the practice of discovery differ. Though this argument is fallacious, its presence, together with the first reason mentioned above, makes it clear that Hall sees abstraction as part of Galileo's theory of discovery.

Hall supports this conclusion by referring to the way in which Galileo in *Il Saggiatore* abstracts the primary qualities from all the properties that bodies have and by referring to the *Dialogue*'s argument against the abstract name *heat* as a possible cause of anything. But this is part of Galileo's practice since it refers to features of his process of reasoning. It need not be decided here whether the practice is one of discovery or of justification. It suffices to point out that Hall is justifying his claim about Galileo's "remarks on the procedure" by appealing to Galileo's actual procedure. That is to say, Hall fails to support his assertion.

The third element of Galileo's "scientific method," according to Hall, is mathematization (pp. 170–74); his discussion of it here is primarily an elucidation and a qualification of the claim. It is not clear, however, what he is elucidating and qualifying. That he introduces mathematization as "an important aid in the process of abstraction" indicates he assumes it to be part of Galileo's theory of method, since I have shown that Hall is talking about Galileo's theory of method in talking about abstraction. The same conclusion is justified by Hall's omission of evidence to support his allegation: he treats Galileo's mathematization as obvious. But what is actually obvious is that Galileo constantly said that mathematics was the proper instrument for understanding nature; whereas it is highly questionable to what extent mathematization was part of his methodological practice, since in the *Dialogue* he did not deal with astronomy in a quantitative way.

Hall's qualification is primarily the qualification of a methodological practice of discovery and partly the qualification of a methodological theory. Hall qualifies the novelty of Galileo's mathematical practice when he states that

> this procedure was not quite new. Optical writers had always treated light-rays and reflecting and refracting surfaces purely geometrically; Archimedes. Galileo's model, had submitted statics to geometry. But no one before had extended the mathematical method of reasoning to the motion of real bodies. [P. 171]

And Hall emphasizes the novelty of Galileo's theory of method when he adds that "nor [had anyone] been so bold as to declare that this method was valid through the whole range of physics: that indeed it was the *only* valid method" (p. 171). For these reasons I conclude that at this point in his own practice Hall treats as interchangeable things that are very different from one another.

The fourth element Hall finds in Galileo's "scientific method" is experimental observation. His claim is here one about either Galileo's theory of method or his practice of discovery, as is shown by Hall's assertion that he is discussing Galileo's answer to the question, "How can the investigator in mathematical physics be sure that his theories are applicable to the real world of experience" (p. 174). But Hall's evidence is that "in the *Discourses,* Galileo verified the law of acceleration by experiments on the inclined plane" (p. 174) and that "in some passages he refers his readers to their experience of reflection of mirrors, of motion of ships under way, of the flow of fluids" (p. 174), which are clearly ideas for experiments if not actual experiments. The above evidence does perhaps indicate that the appeal to observation and actual or imagined experiments constitutes a large part of Galileo's practice of justification. But this is not what Hall is supposed to be concerned with. The third piece of evidence he gives is indeed relevant to one of his concerns, namely, to Galileo's theory of method. Hall quotes the passage from the *Dialogue* in which Galileo speaks of the "sublime wit" of Copernicus, who

> did constantly continue to affirm (being persuaded thereto by reason) that which sensible experiments seemed to contradict; for I cannot cease to wonder that he should constantly persist in saying, that *Venus* revolveth about the Sun, and is more than six times further from us at one time, than at another; and also seemeth always of equal bigness, although it ought to shew forty times bigger when nearest to us, than when farthest off. [Quoted on pp. 174–75]

From this evidence Hall concludes that

> sheer empiricism, therefore, could not uncover physical reality,
> which could only be glimpsed through the alliance of analytical
> reasoning (especially of the mathematical kind), scientific im-
> agination, and cautious experiment always safe-guarded by rea-
> son. [P. 175]

Thus, from the only relevant evidence he gives, Hall draws a conclusion
which is such an overqualified statement of his original claim as to be
actually inconsistent with it. For the original claim was that, by appealing to observation and experiment, the investigator can be sure
that his theories are applicable to reality (p. 174).

My main conclusion is hence not only that Hall's account of Galileo's
"scientific method" is a confused and inconsistent account of four
different things but also that the confusion often prevents him from
achieving what he sets out to prove.

Crombie's Account of Descartes's Scientific Method

Crombie's account of Descartes's "scientific method" consists of two
somewhat overlapping parts, the first given in a discussion of Descartes's contribution to mechanics, and the second in the discussion of
the "philosophy of science and concept of nature in the scientific
revolution." [1] Hence we would expect the first account to treat mainly
of Descartes's methodological practice and the second his theory of
method. To what extent Crombie meets our expectation will become
apparent as the analysis proceeds.

Crombie's first account attributes to Descartes a "method of procedure" which "was in the end fatal" to him:

> But if Descartes can thus be claimed as the first to have given
> expression to the complete principle of inertia, one fundamental
> and in the end fatal distinction between his and Galileo's meth-
> ods of procedure must be emphasized. Whereas Galileo reached
> his incomplete inertial principle as a deduction from a principle
> of conservation of momentum supported by physical reasoning,
> Descartes based his complete principle on an entirely metaphys-
> ical assumption of God's power to conserve movement. [P. 160]

Crombie here tries to contrast Descartes's to Galileo's "method of
procedure." Now, there can be no question that the Cartesian procedure

that Crombie is talking about is a method of grounding, defending, supporting or arguing for something, that is a method of justification; for *based* in this context can only mean "grounded." In speaking of Galileo's "reaching" his principle Crombie could mean his arriving at or discovering it, or he could mean Galileo's concluding or justifying it. Unless we want to say, however, that Crombie is contrasting incommensurables (i.e., a method of justification with a method of discovery), we must conclude that Crombie is talking about Galileo's method of justification. We must be clear about this in order to evaluate properly (1) Crombie's evidence and argument supporting his claim, and (2) the conclusions he wants to draw from that claim. As for the first point, it should be noted that all the evidence that Crombie needs in support of his claim is the correct reproduction of Galileo's and Descartes's arguments justifying their respective principles. It is questionable whether Crombie has faithfully reconstructed Galileo's argument, but we shall neglect this query since we are mainly concerned with Descartes's. As for the Cartesian argument, I think that Crombie states it accurately.

The difficulties begin when we examine the conclusions Crombie wants to draw from his contrast of the Galilean and Cartesian methods of justification. Here we begin to suspect that Crombie may be confounding procedures of justification with procedures of discovery, for what else can he mean in speaking of the "fatal distinction" between the Galilean and Cartesian methods than that the Cartesian method was fatally erroneous? And this thesis seems to be historically as well as theoretically false when made about a method of justification. Historically speaking, Descartes's method of justifying or expounding his view is probably what accounts for the popularity of Cartesianism; even Hall admits this when he says that "the very fact that Descartes wrote as a philosopher [i.e., used the "deductive" method of justification] gave his scientific ideas greater currency. . . ." [2] Theoretically speaking, it can be argued that Descartes's method of arguing from causes or conditions to effects or consequences is the only method of justification which straightforwardly avoids the fallacy of affirming the consequent. Our conclusion must be either that Crombie is mistaken in his implication, or that he must be confounding with methods of discovery the methods of justification that he is actually contrasting. Crombie really has methods of discovery in mind: it is at least partly[3] because of his "method of procedure" that "Descartes completely failed to understand how to measure quantity of motion and thus failed to grasp the essential concept of the conservation of momentum" (p. 161). In fact, in order to be a cause of failure the "method of procedure" in question must be the method of arriving at previously unknown results and not of justifying results already arrived at perhaps by other methods. Therefore, the use to which Crombie wants to put his claim about the contrast between Galileo's and Descartes's methods presupposes that the claim is one concerning their methods of discovery. This presupposition conflicts both with the way that he states his claim and with the evidence he uses to support it, which is evidence about the way Galileo and Descartes justified their principles. Thus there seems to be no way out for Crombie and the difficulties into which he has run must be blamed on his failure to distinguish between methods of discovery and justification.

Crombie's contrast of the Cartesian and Galilean methods of procedure continues:

> In spite of his great contributions to mathematics and to mathematical techniques of physics, Descartes developed his cosmology to a considerable extent largely on non-mathematical lines, and certainly it makes a striking contrast with Galileo's approach to physical problems. [P. 162]

A striking contrast indeed, but hardly more striking than that between Galileo's approach to astronomical problems and Descartes's approach to geometrical problems. And the contrast is hardly more relevant. For Crombie is careful to speak of Galileo's approach to physical problems, meaning, no doubt, problems in mechanics. But this approach contrasts no more to Descartes's approach in cosmology than to Galileo's own attitude in astronomy. Crombie's contrast could hardly be more irrelevant and can only be regarded as a contrast for contrast's sake.

Crombie's next claim does, however, make a prima facie important distinction between Galileo's and Descartes's methods:

> Starting from a background of scholastic physics Galileo achieved his success by eliminating the physical-causal elements from the problem of motion; his approach to dynamics was through kinematics, and . . . his method was to try to solve each individual problem separately, to discover empirically what laws were in fact exhibited in the natural world, before facing the task of reassembling them into a whole. While appreciating

245

> Galileo's individual kinematic descriptions, Descartes found his
> work lacking in a total view of physics and his method of ab-
> straction faulty exactly at the point where Galileo had made it
> effective: its turning away from the problem of physical causes.
> . . . Descartes himself took the opposite course of inquiring be-
> yond the mathematical descriptions into physical causes and the
> nature of things, and of boldly constructing an entire system of
> science. . . . [P. 162]

Crombie clearly tries here to characterize Descartes's method by
opposing it to Galileo's. It is also clear that Crombie regards both as
methods of discovery; otherwise no sense could be made of his implied
causal claims that Galileo's methodic "elimination of the physical-
causal elements from the problem of motion" was a cause, or perhaps
the cause, of his "successes" and that it was "effective." And the "successes" and "effectiveness" of Galileo's methods are for Crombie not
propagandistic success and effectiveness, which Galileo did not really
achieve and which might be caused by a method of exposition or
justification. They are success and effectiveness in the investigation of
nature and discovering its secrets. In brief, only methods of discovery
can be effective in making discoveries.

Limiting ourselves to the discussion on Descartes, our primary interest here, we note that Crombie seems to support his claim by pointing to Descartes's method of exposition or justification, which proceeds
generally and systematically, that is, in a way opposite to the piecemeal
and unsystematic Galilean procedure. But this contrast is irrelevant to
the question of whether the two methods of discovery are either opposite or indeed different.

Crombie also supports his claim with some evidence from two of
Descartes's letters to Mersenne. But this written evidence is hardly appropriate for a claim about his methodological practice; it would be
relevant only to a claim about Descartes's theory of method. At any
rate, the content, insofar as it is relevant, shows how Crombie has
misinterpreted Descartes's method: it shows in fact that Descartes's
opposition to Galileo is only on the method of exposition or justification; and, more importantly, that he agrees with Galileo on the method
of discovery; whereas, on the status of causation and explanation he
seems to have no methodological disagreement with Galileo.

The passages from the two letters are reproduced by Crombie as
follows:

"I will begin this letter with my observations on Galileo's book. I find that in general he philosophises much better than the average, in that he abandons as completely as he can the errors of the schools, and attempts to examine physical matters by methods of mathematics. In this I am in entire agreement with him, and I believe that there is absolutely no other way of discovering the truth. But it seems to me that he suffers greatly from continued digressions, and that he does not stop to explain all that is relevant to each point; which shows that he has not examined them in order, and that, without having considered the first causes of nature, he has merely sought reasons for particular effects; and thus he has built without a foundation." A month later he wrote again: "As to what Galileo has written about the balance and the lever, he explains very well what happens (*quod ita fit*) but not why it happens (*cur ita fit*), as I have done in my *Principles*." [Pp. 162–63]

The first thing to note is that Descartes is talking about a book of Galileo's. Unless he explicitly says otherwise, any comments that he makes about Galileo's method must therefore be regarded as being about Galileo's method of expounding or defending his ideas. Hence, Descartes's objections to Galileo's building without foundations must be objections to his method of justification. Moreover, not only does Descartes explicitly approve Galileo's method of discovery of "examining physical matters by the methods of mathematics"; he also claims to "believe that there is absolutely no other way of discovering the truth." Hence the Galilean method he then objects to is no more a method of discovery than is the method of building with foundations; they are both methods of justification. Finally, on the question of explanation and causation, Descartes shows no awareness of any peculiar Galilean attitude and, consequently, makes no methodological criticism of it; for Galileo "has . . . sought reasons for certain . . . effects," though not for the balance and lever.

This last point also shows that Crombie cannot use the evidence under discussion to support, as he later does by referring to the present evidence, his first claim about Descartes's theory of method. This is the claim that "Descartes criticized Galileo's treatment of motion for providing mathematical description without philosophical basis *and therefore* without explanation" (p. 303, my italics). It is Crombie and not Descartes who believes that lack of philosophical basis implies lack of explanation.

In the rest of his account of Descartes's theory of method, an area in which he could have easily been opposed to Galileo, Crombie is instead more interested in pointing out the similarities between the former and the medievals. This he does, I suppose, in order to add further evidence for his thesis that "a more accurate view of 17th-century science is to regard it as the second phase of an intellectual movement in the West that began when the philosophers of the 13th-century read and digested in Latin translation the great scientific authors of classical Greece and Islam" (p. 110). Had Crombie paid more attention to the distinction between methodological practice and theory of scientific method, he might have been less inclined to accept the "second phase" thesis, which is untenable at the level of methodological practice.

chapter 18

Bridging Internal and External Factors

IN MY EARLIER account of the confusions surrounding the explanation of the rise of modern science I showed primarily the deficiency of the historiographical situation. From another point of view I also showed the unproductiveness of the externalism-internalism and sociology of science-intellectual history of science distinctions both when taken as empirical distinctions and when taken as conceptual ones. I shall now try to clarify the issues, by taking seriously the idea that the general problem is one about *explanation*.

An important part of the problem is causal: What caused the rise of modern science? Another and an even more fundamental part of understanding the rise of modern science is knowing of what the rise of modern science consisted in. A third problem is that of which aspect of the rise is most important, or most contributes to the understanding of it.

To elaborate, beginning with the second problem, if one is to regard the rise of modern science as a historical event, one should think of it as a change, a change from, let us say, medieval science to modern science, or more specifically, as the change from sixteenth to seventeenth-century science. Lest this interpretation seem empty I shall contrast it to two other views. First one might, in true intellectualist-fictionalist fashion, think of the rise of modern science as something like the change from Archimedes and Ptolemy to Copernicus and Galileo. It is true, of course, that, as Kuhn says, the predecessor of Copernicus is Ptolemy and, as Koyré says, the predecessor of Galileo is Archimedes. But this is only true in the sense of the "best predecessors"; it is only in an evaluative—logical and philosophical—sense that those relations hold. I do not accept this alternative view because it makes the rise of modern science look like a logical fiction rather than a historical event. Whereas the historically real predecessors of Copernicus and Galileo are simply the individuals who literally, temporally, lived at a time just before theirs.

One might also view the rise of modern science as the emergence of early modern science independent of the preceding historical situation. This view I reject because in order to appreciate the historical significance or the full meaning of early modern science one must have a contrast with what preceded it. Thus independently of the causal question of whether modern science grew out of or originated from medieval science, knowledge about medieval science is indispensable for understanding what the rise of modern science consisted in. Whether or not one accepts my substantive thesis, the methodological issue should be clear: to understand what the rise of modern science was, is it sufficient to study the activities of the "founders of modern science," or should one also study the activities of their immediate historical predecessors or of only their "worthy" predecessors? There may be no clear-cut answers to these questions, but it is by asking them that one should begin.

So far I have said nothing about which aspect of the rise of modern science should be studied first or most in order to understand it. I have deliberately used the vague terms *modern science* and *medieval science*. I have not referred to the latter as "scholastic natural philosophy" in order not to beg one intellectualist thesis. There are, of course, many aspects of the rise of modern science itself. Independently of the various kinds of "factors" that may have produced it, or of the various kinds of consequences that it had, the rise of modern science is itself a complex phenomenon: it has intellectual, practical, psychological, social, economic, institutional, and, like all other phenomena, physical aspects. In other words, the rise of modern science is at one and the same time the coming into being of certain ideas, forms of behavior, feelings, social interactions and institutions. By contrast, it is not clear whether the rise of modern science has any religious aspect; it is not clear whether the rise of modern science has, in an important sense, a religious character. To say this is not, however, to question that the rise of modern science was in various ways connected with religion. Similarly, it is not clear whether seventeenth-century science, besides interacting with political phenomena, had a properly political aspect.

In the sense that all of the aspects of the rise of modern science are needed for a full understanding of it, they should all be studied. As disagreements may emerge about their relative significance, I suggest the following hierarchy, in decreasing order of importance: intellectual, practical, social, institutional, and psychological aspects. The only two about whose place I am prepared to argue are the intellectual and the practical aspects. No one will probably deny that the intellectual or mental aspect of seventeenth-century science is most important. But the importance of practice has seldom been discussed. Of course experimentation and observation are part of what I mean by "practice"; and historians have considered their role. In general, the practical aspect of the rise of modern science refers to what early modern scientists did as distinct from what they thought. It refers to their actions and behavior, which bring us immediately into the realm of social phenomena, but this need not distract us. The fact is that seventeenth-century scientists not only thought very differently from those of the sixteenth century, but also acted and behaved very differently.

If we start with the view that science is a human activity and remember that actions are no less a part of human life than thoughts, then we will not underestimate the importance of the practical aspects of the rise of modern science. It may be objected that only certain kinds of actions, such as experiments and observations of natural phenomena, are important for an understanding of early modern science. But experiments and observations are only what might be called "scientific forms of behavior," and they are the analog, practically, of what scientific ideas are intellectually or mentally. And if in studying the intellectual aspects of science one does not limit oneself to scientific ideas, but deals with the philosophical and metaphysical as well, why so limit oneself when studying the practice? The answer might be that

251

there is no such thing as philosophical or metaphysical behavior. But the founders of modern science were indubitably engaged in all sorts of activity, so the only relevant question becomes whether there is a unity or interrelatedness in their behavior as there is in their minds. In other words, were their experiments and observations connected with other practices and activities in ways comparable to the way their scientific ideas were related to their philosophical ideas? Actually, since experiments and ideas are not only distinct as practice and theory, but also related in the way that theory and practice are, all of a scientist's practical activities become relatable. To summarize: because of the intimate connection within the life of an individual between theory and practice, between thought and action, the understanding of the rise of modern science depends in an important way on the study of the behavioral aspect of the rise of modern science. The most important activities would be likely to be those relating to technology and to one's profession.

So far all that can be said to a historian of the rise of modern science who, for example, is uninterested in medieval science is that he is "missing" something. The same applies to the historian who, for example, is unwilling or uninterested in studying the practical or social aspects of the rise of modern science. No more than this is missed in the causal context either, as long as all that nonintellectual causes are needed for is to account for nonintellectual aspects of the rise of modern science. This is all that Merton accomplishes:[1] he accounts for the rise in the number of scientists in the seventeenth century in England, a social phenomenon, in terms of the religion, technology, and social conditions prevailing in that country at that time. Analogously, no one should expect intellectual causes to explain nonintellectual aspects of the rise of modern science.

Suppose now we consider the intellectual aspect of the rise of modern science, and let us assume we agree with Koyré's statement that its intellectual content consists of the destruction of the cosmos and the geometrization of space, or the mathematization of nature and of science.[2] Suppose we now ask how and why that content came about. It is clear how someone who asked this question about that aspect, but who had had no previous interest in medieval science, might become interested in medieval science immediately, if it could be shown to him that scholastic natural philosophy encountered a series of intellectual difficulties for the solution of which many changes in its intellectual content had to be made and that their net result was seventeenth-century scientific theories. This explanation does not in fact work, but it is this and only this explanation that would give the intellectual causes of the rise of modern science. Or rather, it is only such an explanation that could do without nonintellectual causes. But Koyré's account has to invoke such nonintellectual causal factors as the translation and publication of Archimedes' works. It is easy enough to see the connection between such a nonintellectual factor and real intellectual developments. But then the question about nonintellectual causes is no longer one of principle, but rather one about which nonintellectual factors can and which cannot be causally connected with intellectual developments. Moreover, if the rise of modern science is as revolutionary a development as Koyré thinks it is, then what becomes hard to understand is that it could have been produced by intellectual factors alone, especially when one considers that the change did not take very long. It becomes easier and not harder to regard it as the result of nonintellectual factors.

Actually, to speak of intellectual versus nonintellectual causes, no less than to speak of internal versus external factors, inclines one to ignore methodological causes. Yet method explanations of the rise of modern science are and have always been very popular, and they allow one to bridge the gap between "internalism" and "externalism." One need not accept the empiricist explanation according to which the rise of modern science was the result of the abandonment of speculation and the adoption of the method of observation and experiment. Nor need one accept the positivist explanation according to which the rise of modern science was the result of abandoning the concern with explanation and 'why' questions and restricting oneself to description and 'how' questions. One could be less simplistic and accept Crombie's view that "the scientific revolution of the 16th and 17th centuries . . . came about by men asking questions within the range of an experimental answer, by limiting their inquiries to physical rather than metaphysical problems, concentrating their attention on accurate observation of the kinds of things that are in the natural world and the correlation of the behavior of one with another rather than on their intrinsic nature, on proximate causes rather than substantial forms, and in particular on those aspects of the physical world which could be expressed in terms of mathematics."[3] Despite its intellectualistic flavor, Koyré's explanation of the rise of modern science has strong methodological elements; the method he emphasizes most is the mathematical approach to the study of nature.

From the present point of view it is not necessary to decide which such explanation is the right one, but to note their common feature: intellectual developments, even scientific ideas, are appropriately explained as being the result of following certain methods, i.e., methodological practices. And the connection between methods and intellectual results, besides being appropriate, is adequate, not because the methods have to lead to the results to which they do, but because it is comprehensible that methods can lead to those intellectual results. It is easy, for example, to see how someone could be led to formulate the time-distance law of falling bodies if he studied the phenomenon and approached it quantitatively and mathematically. But, of course, this is not to say that it is easy to see how someone following the same method would be led to formulate the principle of acceleration. Not just any method can lead to any intellectual result; rather, appropriate methods can lead to appropriate results.

Once the rise of modern science has been reduced to the rise of certain methods (methodological practices) one must find the causes of the rise of those methods to have a complete explanation. Here the more obviously external factors may be considered without inappropriateness or inadequacy. For the procedures constituting the methods may be the result of social, political, economic, or technological factors. Craftsmanship and technology, for example, could have resulted in the rise of the mathematical approach to nature as follows: mathematics and quantitative procedures were being used and used with success in sixteenth-century technology; many people impressed by this success decided to begin using mathematics in natural philosophy. The accuracy of this explanation of the rise of the mathematical method is not in question here, though Zilsel's work tends to support that accuracy.[4] What is in question is its adequacy, about which there can be no question. External factors can be causally connected with methods.

It must be admitted, however, that methods can also be the result of metaphysical ideas. For example, the use of the mathematical method in natural philosophy could be the result of metaphysical mathematicism or Platonism. This is presumably one of Burtt's and Koyré's major theses.[5] And this difference over the cause or causes of a particular agreed-upon method-cause of the new intellectual developments can be taken to be a real disagreement between some externalists and Burtt and Koyré, and their intellectual internalist followers. If a real disagreement, it is hence substantive and not methodological. In fact, both intellectual factors, such as metaphysical ideas, and nonintellectual factors, such as technological or "artistic" practices, could have caused the rise of the new methods.

Real disagreement now becomes possible regarding the explanation of the rise of modern science. The differences are genuine and are not confusions when they pertain to the following two questions: Which specific new methods (methodological practices) were the causes of the new scientific theories of the seventeenth century, and what were the specific causes, admittedly intellectual (metaphysical) *or* nonintellectual (externalist), of the coming into being of these new methods. Now, disagreement is not necessarily an unsatisfactory state of affairs; however it may be, depending on its causes. Or possibly one or more of the disagreeing parties has an untenable position. The genuineness of the disagreement, then, indisputably real at the level of belief, may vanish at the level of evidence.

My causal hypothesis about the conceptual confusion between the contexts of discovery and of justification, and the contexts of theory and of practice, here becomes relevant. When claims about the causes of the new seventeenth-century scientific methods are justified the arguments often reveal a confused and uncritical attitude toward those distinctions.

For example, in his essay "Galileo and Plato," Koyré claims that metaphysical Platonism is the origin of two of Galileo's methods. He alleges Galileo's mathematical approach to be the result of Platonic mathematicism, which he warns us, and rightly so, that we should distinguish sharply from mystical arithmology. In partial support of his claim Koyré refers to other historical works. But he also cites some evidence that, insofar as it is relevant, cannot be accepted because it is verbal evidence, which could only support a claim about Galileo's *theory* of method. It consists of Galileo's own statement that he is a Platonist:

> I know perfectly well that the Pythagoreans had the highest esteem for the science of number and that Plato himself admired the human intellect and believed that it participates in divinity solely because it is able to understand the nature of numbers. And I myself am inclined to make the same judgment.[6]

Koyré also claims Galileo's acceptance of the Platonic doctrine of reminiscence induced him to use the dialog method.[7] He argues:

The allusions to Plato are so numerous in the works of Galileo, and the repeated mention of the Socratic maieutics and of the doctrine of reminiscence, are not superficial ornaments born from his desire to conform to the literary mode inherited from the concern of Renaissance thought with Plato. Nor are they meant to gain for the science the sympathy of the "curious reader," tired and disgusted by the aridity of the Aristotelian scholastics; nor to cloak himself against Aristotle in the authority of his master and rival, Plato. Quite the contrary: they are perfectly serious, and must be taken at their face value. Thus that no-one might have the slightest doubt concerning his philosophical standpoint, Galileo insists:

> SALVIATI: The solution of the question under discussion implies the knowledge of certain truths that are just as well known to you as to me. But, as you do not remember them, you do not see that solution. In this way, without teaching you, because you know them already, but only by recalling them to you, I shall make you solve the problem yourself.
>
> SIMPLICIO: Several times I have been struck by your manner of reasoning, which makes me think that you incline to the opinion of Plato that *nostrum scire sit quoddam reminisci;* pray, free me from the doubt and tell me your own view.
>
> SALVIATI: What I think of this opinion of Plato, I can explain by words, and also by facts. In the arguments so far advanced I have already more than once declared myself by fact. Now I will apply the same method in the inquiry we have in hand, an inquiry which may serve as an example to help you more easily to understand my ideas concerning the acquisition of science. . . .[8]

As for Koyré's argument, first of all the dialog method is a method of justification and not of discovery, and thus Koyré would have the insurmountable problem of connecting that method of justification to Galileo's scientific results. More importantly I think that what Koyré's evidence really shows is that Galileo is in practice justifying Plato's doctrine by the *justification results* that the dialog method yields when put to use. That is, Galileo accepts Plato's doctrine because he accepts the dialog method of justification, not the opposite demanded by Koyré's intellectualist position.

A second example of confusion may be taken from Zilsel, a leading externalist. He argues that Galileo's mathematical quantitative and experimental approach originates from technology, military engineering, and art in order to support his view that on the whole, the rise of the methods of the manual workers to the ranks of academically trained scholars at the end of the sixteenth century is the decisive event in the genesis of science. The upper stratum could contribute logical training, learning, and theoretical interest; the lower stratum added causal spirit, experimentation, measurement, quantitative rules of operations, disregard of school authority, and objective co-operation.[9]

His argument for the origin of Galileo's method is, however, unsatisfactory. In fact, Zilsel uncritically accepts, from the context of theory, such evidence as Galileo's declaration in his letter to Marsili of November 11, 1632 [10] that, in Zilsel's words, "his greatest achievement—the detection of the law of falling bodies, published in the *Discorsi*— developed from a problem of contemporary gunnery." [11] His evidence also includes inappropriate evidence from the context of Galileo's justification of his method: "in his chief work of 1638, the *Discorsi*, the setting of the dialog is the Arsenal of Venice." [12] With no awareness of the difference, he gives in the same passage the following adequate grounds from the context of Galileo's discovery of his method:

When he studied medicine at the University of Pisa in the eighties of the sixteenth century, mathematics was not taught there. He studied mathematics privately with Ostilio Ricci, who had been a teacher at the Accademia del Disegno in Florence, a school founded about seventy years earlier for young artists and artist-engineers. Its founder was the painter Vasari. Both the foundation of this school (1562) and the origin of Galeo's mathematical education show how engineering and its methods gradually rose from the workshops of craftsmen and eventually penetrated the field of academic instruction. As a young professor at Padua (1592–1610), Galileo lectured at the university on mathematics and astronomy and privately on mechanics and engineering. At this time he established workrooms in his house, where craftsmen were his assistants. This was the first "university" laboratory in history. He started his research with studies on pumps, on the regulation of rivers, and on the construction of fortresses. His first printed publication (1606) described a measuring tool for military purposes which he had invented. All his life he liked to visit dockworks and talk with the workmen.[13]

All these claims are about Galileo's practice and are indeed indicative of how he came to adopt his mathematical, experimental approach.

I have already mentioned, in criticizing Crombie's account of

Descartes's scientific method, that those who, like Crombie and Randall,[14] stress the similarity between modern and medieval science in methodological terms simply fail in their aim. Their error is that, whereas the important similarity would be the one in methodological practice, what we find are medieval methodological theories which happen to precede the practice of the seventeenth century. Aside from the problem of how these theories could have been justified, one cannot speak of similarity, but only of conformity of one to the other, a conformity of little consequence unless a causal influence by the medieval methodological theorists on the seventeenth-century methodological practitioners can be established. This has not been done, nor even attempted by the "medievalists." Actually, a true medievalist would better spend his time by trying to find similarities between seventeenth-century science and medieval technology, art, engineering, and crafts, rather than between seventeenth-century science and medieval science. Unawareness or uncritical use of the distinctions between the contexts of theory and practice, is thus the source of the deficiency of even the latest medieval theses. Awareness and critical use of those distinctions promises however what are indeed other forms of the medievalist explanation of the rise of modern science.

To summarize, the main problem concerning the explanation of the rise of modern science is the problem of what caused it. A contextually complete causal account has to involve two steps: (1) explaining the intellectual aspects of the rise of modern science as the result of the rise of certain scientific methodological practices, and (2) explaining the rise of these methods. In principle, the rise of the methods can be accounted for either as the result of metaphysical beliefs or of external factors such as conditions in art and technology. And the distinctions of contexts which I have emphasized are no less important in step (2) than in step (1).

chapter 19

Toward a History of Live Science

The Life of Science

IN EMPHASIZING THE distinctions between the contexts of discovery and of justification, I have also emphasized the much greater importance of the context of discovery for history-of-science explanation. By so doing, I advocate a reform of the historiography of science which would consist in a movement from the literature of science to what might be called the *life of science*. In fact, the context of justification is normally the context of the literary record of the research activity; the method of justification is normally the method of scientific exposition. We may, of course, have to start with this record. But if it is causal explanation of this record that we want, which is to say, if we want to *understand* that record as indeed a record of a research activity, then we have to search for that research activity itself. To some extent this can be done by examining other historical records such as letters, manuscripts, and notes. In general we shall have to delve more into the scientist's "private" life and into his prima facie non-scientific activities. In some cases, when material is objectively or subjectively absent, we might have to resort to pure speculation. This speculation would not be idle because, though perhaps unprovable, it allows us to understand the finished literary record which we possess.

This movement from the literature to the life of science is actually nothing but the second and last phase of a movement that is already well under way. I refer to the current shift in the emphasis from the latest textbook to the historical records themselves. This shift, explicitly noticed by Kuhn,[1] Dijksterhuis,[2] and others, is the result of realizing that the context of a textbook is pedagogical, and thus the logic and methodology found there is the logic and methodology not of science but of the pedagogy of science. The shift is the result of becoming aware, clear, and critical about the distinction between the context of pedagogy and the context of justification. Of course, a logic and methodology of scientific justification—what is found by those historians who study (and stop at) the literary records of the research activity—is better than a logic and methodology of scientific pedagogy. Yet it is not the logic and methodology of science as a live activity, but rather, the logic and methodology of a dead science. For a logic and methodology of live science, we as historians must move to the context of discovery. It is high time that the second shift be made.

The "Logic" of Scientific Discovery

What of the arguments, stemming from Popper, and found in the contemporary literature of the philosophy of science,[3] that there is not and perhaps there cannot be a logic and methodology of scientific discovery? First, it must be emphasized that what they are arguing against is the possibility of a theory of the logic of scientific discovery. That is, they argue that there cannot be general principles of the methodology of scientific discovery. In this I am in agreement. My discussion of the law-coverability anomaly in the concept of explanation of scientific discovery contains some arguments against the possibility of a theoretical logic of discovery. But that is not to say that there are not specific logics of specific scientific discoveries; any satisfactory reason-explanation of a cognition would constitute such a logic; similarly, any satisfactory method-explanation of a scientific success would constitute a specific methodology of a specific discovery. The use of such specific logics and methodologies of specific discoveries is very simple: to explain and thus to understand various aspects of those discoveries. And the usefulness of explanation and understanding is no less than that of a general theory. True, a theory would have a prescriptive use not easily matched. But the specific logics and methodologies too can have a prescriptive use, though not by way of statable prescriptions, but rather by way of intuitive judgments of the similarity of logical and methodological situations. The lack of a theoretical logic of discovery may still be regarded as a defect because of the possible predictive use of the theory. Here indeed the specific logics of specific discoveries are impotent, but for this we ought to be thankful since a predictive theory of the logic of discovery would actually imply the end of the growth of human knowledge by way of the divinization of the mind of man, an unwelcome prospect.

Second, in the sense in which those philosophies reject a logic of scientific discovery, they accept a logic of scientific justification. With this I do not agree. I believe that a theoretical logic of scientific justification is no more tenable than one of scientific discovery. My arguments against philosophies of the history of science support, in part, my belief.

Third, the same philosophers do not reject the possibility of a logic of discovery; rather, what they do is to equate the logic of discovery with what they take to be logic of justification. Popper's arguments against genetic explanations (arguments based on the idea that the origin of an idea has no logical importance) become relevant here. To this it may be replied that the genesis of an idea constitutes the historical explanation of its emergence; hence if we are interested in explanation and understanding we cannot do without genetic accounts. But should one be interested in genesis? I have argued above that one must be in order to come to grips with the life of science. Here the Popperian might reply that his emphasis on what one does with an idea after he has it *is* the life of science, or at least the most important aspect of it. To this it may be said that one's attitude to an idea after one has it is no less a private affair than the way in which one had the idea in the first place.

At any rate, the Popperians cannot escape the spirit of my distinction between the contexts of discovery and of justification—or the one between the contexts of theory and practice, for that matter. In fact, the notion of what one does with an idea is ambiguous: it may refer to what one does in the context of the literary work or to what one did in real life or research. And the Popperians too have succumbed to the habit of examining primarily what a scientist does with an idea in the context of his literary exposition. It is well known that one of the sources of Popper's hypothetico-deductive logic of discovery-justification is Galileo's "work." The work most used is the *Discourses on Two New Sciences*, a *book*. Nor do the Popperians escape the temptation of being unmindful of the distinction between the contexts of theory and practice; witness their frequent references to Einstein's own reflections on the logic and methodology of his discoveries. For example, they admit their reliance on Einstein's theory of method as set forth in his lecture "On the Method of Theoretical Physics." [4] All the Popperians have done then is to emphasize different aspects of the literature of science and of the words of scientists.

History of Science and History of Philosophy

Some of my criticism of Koyré is a reaction against the idea and practice of modeling the history of science on the history of philosophy, an ideal which Hall [5] attributes to the intellectual historian of science. Actually the modeling was intended to apply only regarding the emphasis on intellectual aspects; it certainly was not meant to imply that the history of science should regress to being inductivist chronologies of the latest textbook, a description which would still fit the histories of philosophy published in English-speaking countries. It may seem hard to account for this inductivist character of histories of philosophies when most historians of philosophy allegedly believe that in philosophy there is not an analog of the scientific text. Presumably they see philosophy as having only problems and no solutions. Yet, a history of philosophy does have that character when it is a chronology of philosophical problems.

Although the emphasis on intellectual aspects in the history of science is in itself valuable, two great dangers attend this way of imitating the history of philosophy. The first, which seems to have become actualized, is that the intellectual aspects of historical developments will be studied to the exclusion of other aspects, such as the practical ones relating to the actions of individuals. The emphasis on intellectual matters, valuable in itself, yields diminishing returns when pursued without at least a minimum emphasis (I will not say a *comparable* emphasis) on the practical, i.e., behavioral, aspects.

The second danger is that the emphasis on the intellectual aspects may turn out to be no more than an emphasis on the context of justification, that is, of the finished literary record. To the detriment of the history-of-science explanation and understanding, this is what seems actually to have happened. And it should hardly be a surprise, considering that practically all of what is called history of philosophy completely ignores the context of discovery and concentrates almost exclusively on textual analysis. It is, of course, beyond the scope of my investigation to prove that the neglect of the context of philosophical discovery by the historian of philosophy leads him to what he himself would regard as undesirable consequences. Yet, in order to pursue his own objectives more efficiently he should be more concerned than he is with the intellectual causes of philosophical ideas.

The Autonomy of History of Science

In very general terms, I argue for the autonomy of history of science, or at least of history-of-science explanation. My solution to the problem of explanation in the historiography of science tends to dilute the historian's dependence on science itself. It frees history of science from its dependence on what scientists *say* about their practice; it is better to disregard their reflections on their practice than to be uncritical of them. The historian of science is also freed from science in the sense that he does not need to know much contemporary science. In fact, if, as I suggested, the frequent divergence between the logic of discovery and the logic of justification is the effect of the growth of knowledge, then knowledge of later science may prevent the historian, or make it difficult for him, to find the specific logic of discovery, which is what he needs for explanation. In other words, it seems that the better one knows the logic of justification, the harder it is to reconstruct the logic of discovery.

Toward a Historicist Philosophy of Science

My thesis lays the foundation for a *historicist philosophy* or philosophical history of science, that is to say, a historical approach to the nature of science. Such historicist philosophy of science would not be a philosophy of the history of science, like the Popperian and Kuhnian doctrines I have criticized. It would be neither a theoretical history of science nor a comprehensive history of science which would attempt to classify and organize all historical facts. Instead, it would be history, in my interpretation of Croce's sense of history, since it would be concerned with understanding; and it would be an explanatory activity, in my interpretation of Scriven's sense of explanation. It would be philosophy in that it would be concerned with criticism of present and past science and scientists. It would thus be concerned with the nature of science, in order to ground its criticism. It would combine history and philosophy because the *nature* of science would be taken to be its *history*.

notes

Chapter 1

1. The term *historiography of science* will always refer to the historical study of the development of science, i.e., the discipline of the historian of science. Whereas the term *history of science* will sometimes refer to the development of science and sometimes to the discipline, the context will usually indicate which is meant. Moreover, the use or the omission of the definite article *the* just before the term *history of science* has been found to offer a natural way of distinguishing between the development and the discipline, respectively.

2. Joseph Agassi, *Towards an Historiography of Science,* Supplement 2, *History and Theory.* (The Hague: Mouton, 1963).

3. Thomas S. Kuhn, "Review of Agassi's 'Towards an Historiography of Science'," *British Journal for the Philosophy of Science,* 17:257, November 1966.

4. Roger Hahn, "Reflections on the History of Science," *Journal of the History of Philosophy,* 3:239, October 1965.

5. W. A. Smeaton, "Review of Agassi," *Annals of Science,* 18:125–29, 1962.

6. Ernest Nagel, *The Structure of Science* (N.Y.: Harcourt, Brace and World, 1961).

7. Agassi, *Historiography.*

8. Carl G. Hempel, "Aspects of Scientific Explanation," in *Aspects of Scientific Explanation* (N.Y.: Free Press, 1965).

9. Karl Popper, *The Poverty of Historicism* (N.Y.: Harper, 1957).

10. Agassi, *Historiography.*

11. The term *history-of-science explanation* will always refer to the explanation of facts or events of the development of science, such as a historian of science may be expected to give.

12. Arguments in support of the autonomy of ordinary historical explanation can be found in Michael Scriven, "Truisms as the Grounds for Historical Explanation," *Theories of History,* ed. P. Gardiner (Glencoe, Ill.: Free Press. 1959), pp. 443–75; and "New Issues in the Logic of Explanation," *Philosophy and History,* ed. Sidney Hook (N.Y.: New York University Pr., 1963), pp. 339–61.

Chapter 2

1. Carl Hempel, *Aspects,* p. 367.

2. Ibid.; and "Reasons and Covering Laws in Historical Explanation," *Phi-*

losophy and History, ed. Sidney Hook (N.Y.: New York University Pr., 1963), pp. 143–63.

3. Hempel, *Aspects,* pp. 367–68.

4. Hempel, *Philosophy and History,* p. 146.

5. Ibid.

6. Hempel, *Aspects,* pp. 367–68.

7. Michael Scriven, "Explanation and Prediction in Evolutionary Theory," *Science,* 130:477, 1959.

8. Ibid., p. 478.

9. Ibid.

10. Hempel, *Aspects,* p. 370.

11. N. R. Hanson, "On the Symmetry Between Explanation and Prediction," *Philosophical Review,* 68:349–59, 1959; and *The Concept of the Positron* (Cambridge: Cambridge University Pr., 1963), chapter 2.

12. Hempel, *Aspects,* p. 407.

13. Mentioned in Hempel, *Aspects,* p. 407.

14. Hempel, ibid., p. 407.

15. Hempel, ibid., p. 408.

16. P. K. Feyerabend, "Review of Hanson," *Philosophical Review,* 73:264–66, 1964.

17. Michael Scriven, "Truisms as the Grounds for Historical Explanations," *Theories of History,* ed. P. Gardiner (Glencoe, Ill.: Free Press, 1959), p. 469.

18. Ibid.

19. Hempel, *Aspects,* p. 371.

20. Ibid.

21. Scriven, "Explanation and Prediction," p. 480.

22. Scriven, "Truisms," p. 469.

23. See, for example, Scriven, "Explanation and Prediction," p. 480.

24. Arthur Koestler, *The Sleepwalkers* (N.Y.: Grosset and Dunlap, 1959), p. 288.

25. Hempel, *Aspects,* pp. 366, 367, 368.

26. This explanation is found in Agassi, *Historiography,* p. 58.

27. Hempel, *Aspects,* p. 471.

Chapter 3

1. Michael Scriven, "Truisms," pp. 443–75.

2. Ibid.

3. Karl Popper, *The Logic of Scientific Discovery* (N.Y.: Harper & Row, 1959), p. 253.

4. For some relevant arguments, see Scriven, "Truisms"; and "Explanations, Predictions, and Laws," *Minnesota Studies in the Philosophy of Science,* vol. 3, ed. H. Feigl and G. Maxwell (Minneapolis: University of Minnestoa Pr., 1962), pp. 170–230.

Chapter 4

1. This point has been made by Scriven in the literature.

2. Hempel, *Aspects,* p. 367.

3. This example has been discussed by both Koestler and Agassi in contexts which are somewhat related to the present one but different. More specifically Koestler discusses this discovery as an example of what he calls "bisociation" to support his own theory of creativity; see his *Insight and Outlook* (N.Y.: Macmillan, 1949), pp. 251–55. Agassi discusses the discovery as an example of the methodological problem of wisdom after the event and is primarily concerned with supporting his idea that the historian should search for *intellectual* obstacles to scientific discoveries, e.g., the scientific theories accepted at the time; see pp. 51–54 of his *Historiography.*

4. Agassi, *Historiography.*

5. Henry Guerlac, *Lavoisier—The Crucial Year* (Ithaca, N.Y.: Cornell University Pr., 1961).

6. Alexandre Koyré, "La Loi de la Chute des Corps. Descartes et Galileo," *Etudes Galiléennes,* vol. 2 (Paris: Hermann, 1939). Reprinted in Alexandre Koyré, *Etudes Galiléennes* (Paris: Hermann, 1966), pp. 83–158.

Chapter 5

1. The original French version of the last section of Lavoisier's August memorandum is reproduced as follows on page 214 of H. Guerlac's *Lavoisier—The Crucial Year* (Ithaca: Cornell University Pr., 1961). The words and phrases crossed out by Lavoisier himself are placed here, as by Guerlac, in parentheses; Lavoisier's own emendations and additions are here printed, as by Guerlac, in italics.

SUR L'AIR FIXE, ou plutot,

Sur l'air Contenu dans les Corps.

Il paroit Constant que (la pluspart) l'air entre dans la Composition de la plus part des Mineraux, *des Metaux meme Et en tres grande abondance.* aucun chimiste Cependant n'a fait encore entrer l'air dans la definition ni des Metaux ni d'aucun corps mineral. une effervescence n'est autre chose qu'un degagement Subit de l'air qui etoit en quelque façon dissout dans chacun des Corps que l'on Combine.

Ce degagement a lieu toutes les fois qu'il entre moins d'air dans la Combinaison du nouveau Composé qu'il n'entroit dans chacun des deux Corps qui entrent dans la Combinaison. (on ne Suivra pas ces Vues plus loin. elles sont le Sujet d'un travail deja fort avancé meme en partie Redigé) *Ces vues suivies et approffondies pourroient Conduire a une theorie interessante quon a meme deja ebauchée,* mais ce qui doit ici fixer l'attention, c'est que la plus part des Metaux ne font plus d'effervescence lors qu'ils ont ete tenus longtemps au feu du Miroir ardent. Sans doute que le degré de Chaleur qu'ils y eprouvent leur enleve l'air qui entroit dans leur Combinaison. ce qui est tres particulier, c'est que les metaux dans cet etat ne Sont plus malleables Et qu'ils Sont presqu' indissolubles dans les acides. cette observation qui a encore besoin de Confirmation peut fournir une ample matierre a observations Et a Reflexions.

il Seroit bien a desirer qu'on put appliquer au Verre ardent l'appareil de M. halles [sic] pour mesurer la quantité d'air *produitte ou absorbée dans chaque operation,* mais on craint que les difficultés que presentent ce genre d'experience ne Soit insurmontable au Verre ardent.

Chapter 6

1. Alexandre Koyré, "La Loi de la Chute des Corps. Descartes et Galileo," *Etudes Galiléennes* (Paris: Hermann, 1966), pp. 84–85, my italics; the translation from the French is my own here and in all other quotations from this work. This 1966 edition of *Etudes Galiléennes* consists of a collection of three works previously published in 1939 as three separate volumes; "La Loi de la Chute des Corps. Descartes et Galileo" constitutes the entire second volume of the 1939 edition. This second volume has two sets of page numbers; one set can be obtained from my references to the 1966 edition by subtracting 82 and prefixing the Roman numeral two.

2. Quoted in French by Koyré, *Etudes Galiléennes,* pp. 103–105, from the Italian text found in Galileo Galilei, "Frammenti Attenenti ai Discorsi," *Opere,* vol. 8, ed. A. Favaro (Florence, 1890–1909), p. 373, my italics.

3. Quoted in French by Koyré, *Etudes Galiléennes,* pp. 113–14, from the Latin text in René Descartes, "Cogitationes Privatae," *Oeuvres,* vol. 10, ed. Adam and Tannery, p. 219.

4. Quoted in French by Koyré, *Etudes Galiléennes,* pp. 116–17, from the Latin text in René Descartes, "Physico-Mathematica," *Oeuvres,* vol. 10, pp. 75–76.

5. Descartes, "Cogitationes Privatae," *Oeuvres,* vol. 10, p. 219; quoted by Koyré, p. 114, n. 1.

6. Koyré, *Etudes Galiléennes,* pp. 147–48; the Galilean passage is from Galileo Galilei, *Dialogo (Opere,* vol. 7), pp. 255–56.

7. Galileo Galilei, *Dialogue Concerning the Two Chief World Systems,* trans. S. Drake (Berkeley: University of California Pr., 1967), pp. 229–30.

8. Galileo, *Dialogue,* trans. S. Drake, p. 229, my italics. The original Italian reads: "E però, se il mobile che cadendo si è servito de i gradi di velocità accellerata, conforme al triangolo ABC, ha passato in tanto tempo un tale

spazio, è ben ragionevole e probabile che servendosi delle velocità uniformi e rispondendo al parallelogrammo, passi con moto equabile nel medesimo tempo spazio doppio al passato dal moto accellerato." Cf. Galileo, *Dialogo* (*Opere*, vol. 7), p. 256.

9. Galileo Galilei, *Dialogues Concerning Two New Sciences*, trans. H. Crew and A. De Salvio (N.Y.: Dover, n.d.), p. 160, my italics.

Chapter 7

1. George Basalla, ed., *The Rise of Modern Science: External or Internal Factors?* (Lexington: Heath, 1968).

2. Basalla, *Rise of Modern Science*, p. vii.

3. Ibid., pp. xv–xvi.

4. Thomas S. Kuhn, "Science: The History of Science," *International Encyclopedia of the Social Sciences*, 14:74–83.

5. Kuhn, "Science," p. 76, my italics.

6. Basalla, *Rise of Modern Science*, pp. vii–xiv.

7. A. R. Hall, "Merton Revisited, or Science and Society in the Seventeenth Century," *History of Science*, 2:1–15, 1963.

8. Basalla, *Rise of Modern Science*, p. x.

9. B. M. Hessen, "The Social and Economic Roots of Newton's Principia," *Science at the Cross Roads* (London: Kniga, 1931), p. 36.

10. Ibid., p. 40.

11. Alexandre Koyré, "Newton and Descartes," *Newtonian Studies* (London: Chapman & Hall, 1965), pp. 82–83.

12. G. N. Clark, *Science and Social Welfare in the Age of Newton* (Oxford: Clarendon Pr., 1949), pp. 86–89.

13. A. R. Hall, "Merton Revisited," *History of Science*, 2:1–15, 1963; also in Basalla, ed., *Rise of Modern Science*, pp. 89 06.

14. Hall, "Merton Revisited," p. 14; in Basalla, p. 96.

15. Hall, "Merton Revisited," p. 6.

16. See, for example, Scriven, "New Issues in the Logic of Explanation," *Philosophy and History*, pp. 339–61.

17. Koyré, "Newton and Descartes," pp. 53–114.

18. Hall, "Merton Revisited," p. 8.

19. Ibid., pp. 10–11.

20. Ibid., p. 1.

Chapter 8

1. Karl Popper, "The Aim of Science," *Ratio*, 1:24–35, December 1967.

Chapter 9

1. Agassi might object to our plan on the grounds that it is not really fair to analyze his claims by means of the model because they are claims in the field of the history of the historiography of science and the model need not apply here since this discipline belongs to a higher level epistemological category than other more common ones.

To this argument one could reply that, besides *saying* that the deductive model applies to mathematics, physics, history, and sociology, Agassi has applied it to the history of science, which by his own counting would be on a level different from and higher than science. Therefore, he has committed himself, so to speak, not only to the generalization of application from one field of first level to all fields of first level, in a horizontal direction, but also to its extension from level to level, in a vertical direction.

Moreover it is clear that Popper's theory of explanation is an essential part of both his philosophy of science and of Agassi's philosophy of the historiography of science, since these philosophies claim, respectively, that scientific theories are explanations of facts and that philosophies of science are explanations of facts of the development of science. Both of these claims would therefore be empty or ambiguous unless the notion of explanation is explicated. Therefore to apply the "so-called deductive model" to the explanations put forth by Agassi is, in this respect, simply an exercise for determining the inner consistency of Agassi's philosophy of the historiography of science. Thus the fairness of applying the model to Agassi's explanations is beyond dispute.

2. Karl Popper, *The Logic of Scientific Discovery* (N.Y.: Harper & Row, 1959), pp. 59–60.

3. Ibid., p. 60, n. 1.

4. Ibid., pp. 59–60.

Chapter 10

1. Agassi, *Historiography*, p. viii.

2. Ibid., p. 77.

3. Ibid.

4. Ibid., p. v.

5. Cf. Henry Guerlac, "The Continental Reputation of Stephen Hales," *Archives Internationales des Sciences*, 4:393–404, 1951; and "A Lost Memoir of Lavoisier," *Isis*, 50:125–29, 1959.

6. Alexandre Koyré, "Galileo and the Scientific Revolution of the Seventeenth Century," and "Galileo and Plato," *Metaphysics and Measurement* (Cambridge: Harvard University Pr., 1968), pp. 1–43, especially pp. 2, 3, 19–21, 22, 38, 39, 43; also "The Significance of the Newtonian Synthesis,"

Newtonian Studies (London: Chapman & Hall, 1965), pp. 3–24, especially pp. 4–5, 6, 8.

7. Koyré, "Galileo and Plato," p. 40, n. 2.

8. C. C. Gillespie, *The Edge of Objectivity* (Princeton: Princeton University Pr., 1960).

9. Alexandre Koyré, "Galileo and the Scientific Revolution of the Seventeenth Century," *Philosophical Review*, 52:333–48, 1943; and "Galileo and Plato," *Journal of the History of Ideas*, 4:400–428, 1943; both reprinted in *Metaphysics and Measurement*.

10. Alexandre Koyré, "The Significance of the Newtonian Synthesis," Lecture, University of Chicago, November 3, 1948; reprinted in A. Koyré, *Newtonian Studies* (London: Chapman & Hall, 1965), pp. 3–24.

Chapter 11

1. C. C. Gillispie, *The Edge of Objectivity* (Princton, N.J.: Princeton University Pr., 1960), pp. 3–7.

2. Ibid., pp. 523, 527.

3. Thomas S. Kuhn, *The Copernican Revolution* (Cambridge: Harvard University Pr., 1957).

4. Michael Scriven, "Explanations, Predictions, and Laws," pp. 170–72; and Michael Scriven, "Science: The Philosophy of Science," *International Encyclopedia of the Social Sciences*, 14:83–92.

5. I have to refer to norms and propriety because no one would want to follow, e.g., R. H. Dicke and J. P. Wittke when on pp. 5 and 367 of their *Introduction to Quantum Mechanics* (Reading: Addison-Wesley, 1960), they call the Rayleigh-Jeans *incorrect* formula of blackbody radiation a law.

6. Michael Scriven, "The Key Property of Physical Laws—Inaccuracy," *Current Issues in the Philosophy of Science*, ed. H. Feigl and G. Maxwell (N.Y.: Holt, Rinehart & Winston, 1961), p. 100.

7. Alexandre Koyré, *Etudes Galiléennes*, p. 87.

Chapter 12

1. Agassi, *Historiography*, pp. 23–28.

2. Thomas S. Kuhn, *The Structure of Scientific Revolutions* (Chicago: University of Chicago Pr., 1970).

3. Agassi, *Historiography*, p. 64. Also P. K. Feyerabend, "Explanation, Reduction, and Empiricism," *Minnesota Studies in the Philosophy of Science*, vol. 3, ed. H. Feigl and G. Maxwell (Minneapolis: University of Minnesota Pr., 1962), pp. 28–97; "Reply to Criticism" in *Boston Studies in the Philosophy of Science*, vol. 2, ed. R. S. Cohen and M. W. Wartofsky (N.Y.: Humanities Pr., 1965), pp. 223–61; "Review of Nagel's *The Structure of*

Science," British Journal for the Philosophy of Science, 17:237–49, November 1966.

4. F. Cajori, ed., *Sir Isaac Newton's Mathematical Principles of Natural Philosophy and His System of the World* (Berkeley: University of California Pr., 1934), p. 21.

5. Cajori, ibid., p. 21.

6. Galileo Galilei, *Dialogues Concerning Two New Sciences,* p. 174.

7. Ibid., p. 245.

8. Ibid., pp. 244–52.

9. Feyerabend, "Reply to Criticism," pp. 223–25.

10. Karl Popper, *The Logic of Scientific Discovery* (N.Y.: Harper & Row, 1959), p. 53.

11. T. S. Kuhn, *The Structure of Scientific Revolutions.*

12. Many objections could be made to the details of Kuhn's line of argument. For example, we could ask whether the set of properties of scientific communities that Kuhn lists (p. 168), is (1) sufficient to characterize as scientific all of the communities of the physical sciences, (2) consistent with the rest of his book since the condition that sciences must study nature necessarily excludes the social sciences, history, philosophy, and mathematics, and since he speaks of these as sciences or pre-sciences.

13. Alexandre Koyré, "Newton and Descartes," pp. 53–114.

Chapter 13

1. C. C. Gillispie, "Review of Agassi's 'Towards an Historiography of Science'," *Isis,* 55:97, 1964.

2. Thomas S. Kuhn, "Review of Agassi's 'Towards an Historiography of Science'," *British Journal for the Philosophy of Science,* 17:257, November 1966.

3. Benedetto Croce, *History as the Story of Liberty* (London: George Allen and Unwin, 1941), trans. S. Sprigge, pp. 17–18.

4. Roger Hahn, "Reflections on the History of Science," *Journal of the History of Philosophy,* 3:242, October 1965.

5. Benedetto Croce, *Theory and History of Historiography* (London: George G. Harrap & Co., 1921), trans. D. Ainslie, pp. 25–26; see also B. Croce, "History and Chronicle," *Theories of History,* ed. P. Gardiner (Glencoe, Ill.: Free Press, 1959), pp. 226–33.

6. Benedetto Croce, *Teoria e Storia della Storiografia* (Bari: Laterza, 1966). This is my translation of the sentence following the one referred to in note 5 above.

7. Croce, *History as the Story of Liberty,* p. 16.

8. Morton White, "The Logic of Historical Narration," *Philosophy and History,* ed. Sidney Hook (N.Y.: New York University Pr., 1963), p. 6.

9. Michael Scriven, "Explanations, Predictions, and Laws," pp. 170–230; and "Truisms," pp. 443–75.

10. For a discussion of such a continuum, see Scriven, "Causes, Connections, and Conditions in History," *Philosophical Analysis and History*, ed. W. Dray (N.Y.: Harper, 1966), pp. 238–64.

11. Scriven, "Explanations, Predictions, and Laws" and "Truisms."

12. Croce, *Theory and History of Historiography*, pp. 19–20; I have amended Mr. Ainslie's translation in a few places; see also "History and Chronicle," pp. 226–33.

Chapter 14

1. Karl Popper, *The Logic of Scientific Discovery* (N.Y.: Harper & Row, 1959), p. 59.

2. Michael Scriven, "Truisms," pp. 443–75; and Scriven, "Explanations, Predictions, and Laws," pp. 170–230. This idea constitutes the central tenet of Michael Scriven's theory of explanation and has been upheld and defended by him against confusions on the part of positivist philosophers of science. If the above mentioned logical feature is indeed present in Guerlac's and Koyré's works, the philosophical adherents to Scriven's idea will regard it as a reason for treating those works as explanations. On the other hand, historians of science, who probably will first accept or judge those works as explanations, will thereby have reason to accept Scriven's analysis of explanation, on grounds that this analysis corresponds to their practice and intuition about explanation.

3. Agassi, *Historiography*, pp. v, 4, 77.

4. Thomas S. Kuhn, "Science: The History of Science," *International Encyclopedia of the Social Sciences*, 14:81–2.

5. I. B. Cohen, "History of Science as an Academic Discipline: Discussion," *Scientific Change*, ed. A. C. Crombie. *Symposium on the History of Science*, Oxford, 1961. (London: Heineman, 1963), pp. 769–80.

6. Agassi, *Historiography*, p. 77.

7. Peter Doig, *A Concise History of Astronomy* (London: Chapman & Hall, 1950).

8. Kuhn, *Structure of Scientific Revolutions*, p. 116.

9. Ibid.

10. Ibid., p. 115.

11. Ibid., p. 116.

12. Doig, *Concise History*, pp. 111–16.

13. Croce, "History and Chronicle," pp. 226–33; see also Croce, *Theory and History of Historiography*, especially chapter 1.

14. Croce, *Theory and History of Historiography*, pp. 26, 168.

15. Scriven, "Truisms" and "Explanations, Predictions, and Laws."

Chapter 15

1. Scriven, "Truisms," pp. 443–75.

2. These terms have been used by Popper, and similar ones by Hempel, with a different meaning from mine; hence my use of those terms should not be confused with theirs. I shall restrict my use of them to situations where it is stylistically necessary.

3. Henry Guerlac, *Lavoisier—The Crucial Year* (Ithaca, N.Y.: Cornell University Pr., 1961), p. 34.

4. A. R. Hall, *The Scientific Revolution: 1500–1800* (Boston: Beacon Pr., 1962), p. 340.

5. A. C. Crombie, *Medieval and Early Modern Science* (Garden City, N.Y.: Doubleday, 1959), 2:162.

Chapter 16

1. Koyré, *Etudes Galiléennes* (Paris: Hermann, 1966), p. 137.

Chapter 17

1. A. C. Crombie, *Medieval and Early Modern Science*, 2:159–65, 303–308. Whether one is interested in Descartes's method or in the historiographical problems connected with giving an historical account of that method, an excellent secondary source to use is either Hall or Crombie. Most other accounts are limited, either because they are too short, e.g., E. J. Dijksterhuis, *The Mechanization of the World Picture* (Oxford: Clarendon Pr., 1961), pp. 403–18; and E. A. Burtt, *The Metaphysical Foundations of Modern Physical Science* (Garden City, N.Y.: Doubleday, 1954), pp. 105–10; or because they are presented from a limited point of view, e.g., M. B. Hesse, *Forces and Fields* (London: Thomas Nelson and Sons, 1961), pp. 108–12; or because they consist of summaries of Cartesian works on method, e.g., J. F. Scott, *The Scientific Work of René Descartes* (London: Taylor and Francis, 1952).

2. Hall, *Scientific Revolution*, p. 182.

3. I say "partly" because, according to Crombie, another cause of the Cartesian failure was his excessively geometrical outlook.

Chapter 18

1. R. K. Merton, "Science, Technology and Society in Seventeenth Century England," *Osiris*, 4:361–632, 1938.

2. Alexandre Koyré, "The Significance of the Newtonian Synthesis," pp. 6–7.

3. A. C. Crombie, *Medieval and Early Modern Science,* 2:121.

4. See, for example, E. Zilsel, "The Sociological Roots of Science," *The American Journal of Sociology,* 47:544–62, January 1942.

5. A. Koyré, "Galileo and Plato," and E. A. Burtt, *The Metaphysical Foundations of Modern Science* (Garden City, N.Y.: Doubleday, 1954).

6. Quoted in Koyré, "Galileo and Plato," pp. 40–41 from Galileo Galilei, *Dialogo* (vol. 7 of *Opere*), p. 35.

7. Koyré, "Galileo and Plato," p. 42.

8. Ibid., quoted from Galileo, *Dialogo,* p. 217.

9. E. Zilsel, "The Sociological Roots of Science," 47:558.

10. Galileo Galilei, *Opere,* 14: 386.

11. Zilsel, "Sociological Roots," p. 556.

12. Ibid.

13. Ibid., pp. 555–56.

14. A. C. Crombie, *Medieval and Early Modern Science,* vol. 2; and J. H. Randall, "The Development of Scientific Method in the School of Padua," *Journal of the History of Ideas,* 1:177–206, 1940.

Chapter 19

1. Kuhn, *Structure of Scientific Revolutions,* p. 1.

2. E. J. Dijksterhuis, *The Mechanization of the World Picture* (Oxford: Clarendon Pr., 1961), pp. 344–45.

3. P. K. Feyerabend, "Comments on Hanson's 'Is There a Logic of Discovery?'" *Current Issues in the Philosophy of Science,* ed. H. Feigl and G. Maxwell (N.Y.: Holt, Rinehart & Winston, 1961), pp. 35–39.

4. A. Einstein, "On the Method of Theoretical Physics," *Ideas and Opinions* (N.Y.: Crown, 1954), pp. 270–76.

5. A. R. Hall, "Merton Revisited," p. 93.

bibliography

BOOKS

Agassi, Joseph. *Towards an Historiography of Science.* Supplement 2, *History and Theory.* Hague: Mouton, 1963.

Basalla, George, ed. *The Rise of Modern Science: External or Internal Factors?* Lexington, Mass.: Heath, 1968.

Burtt, Edwin A. *The Metaphysical Foundations of Modern Physical Science.* Garden City, N.Y.: Doubleday, 1954.

Clark, George N. *Science and Social Welfare in the Age of Newton.* Oxford: Clarendon Press, 1949.

Cohen, Robert, and Marx W. Wartofsky, eds. *Boston Studies in the Philosophy of Science,* vol. 2. N.Y.: Humanities Press, 1965.

Croce, Benedetto. *History as the Story of Liberty.* Translated by Sylvia Sprigge. London: G. Allen & Unwin, 1941.

————. *Teoria e Storia della Storiografia.* 2nd revised edition. Bari: Giuseppe Laterza & Figli, 1920.

————. *Theory and History of Historiography.* Translated by Douglas Ainslie. London: George G. Harrap & Co., 1921.

Crombie, Alistair C. *Medieval and Early Modern Science,* vol. 2. Garden City, N.Y.: Doubleday, 1959,

————, ed. *Scientific Change.* Symposium on the History of Science, Oxford, 1961. London: Heineman, 1963.

Descartes, René. *Oeuvres,* 12 vols. Edited by Charles Adam and Paul Tannery. Paris: Leopold Cerf, 1897.

Dijksterhuis, Eduard J. *The Mechanization of the World Picture.* Translated by C. Dikshoorn. Oxford: Clarendon Press, 1961.

Doig, Peter. *A Concise History of Astronomy.* London: Chapman & Hall, 1950.

Dray, William, ed. *Philosophical Analysis and History.* N.Y.: Harper & Row, 1966.

Feigl, Herbert, and Grover Maxwell, eds. *Scientific Explanation, Space, and Time.* In *Minnesota Studies in the Philosophy of Science,* vol. 3. Minneapolis: University of Minnestota Press, 1962.

Galilei, Galileo. *Dialogue Concerning the Two Chief World Systems.* Translated by S. Drake. Berkeley: University of California Press, 1953.

————. *Dialogues Concerning Two New Sciences.* Translated by H. Crew and A. De Salvio. N.Y.: Dover, n.d.

————. *Opere,* 20 vols. Edited by Antonio Favaro. Florence: G. Barbera, 1890–1909.

Gardiner, Patrick, ed. *Theories of History*. Glenco, Ill.: Free Press, 1959.

Gillispie, Charles C. *The Edge of Objectivity: An Essay in the History of Scientific Ideas*. Princeton, N.J.: Princeton University Press, 1960.

Guerlac, Henry. *Lavoisier—The Crucial Year*. Ithaca, N.Y.: Cornell University Press, 1961.

Hall, A. Rupert. *The Scientific Revolution: 1500–1800*. Boston: Beacon Press, 1962.

Hempel, Carl G. *Aspects of Scientific Explanation*. N.Y.: Free Press, 1965.

Hook, Sidney, ed. *Philosophy and History*. N.Y.: New York University Press, 1963.

Koestler, Arthur. *The Sleepwalkers*. N.Y.: Grosset and Dunlap, 1959.

Koyré, Alexandre. *Etudes Galiléennes*. Paris: Hermann & Co., 1966.

————. *Metaphysics and Measurement*. Cambridge, Mass.: Harvard University Press, 1968.

————. *Newtonian Studies*. London: Chapman & Hall, 1965.

Kuhn, Thomas S. *The Copernican Revolution*. Cambridge: Harvard University Press, 1957.

————. *The Structure of Scientific Revolutions*. Chicago: University of Chicago Press, 1970.

More, Louis T. *Isaac Newton*. N.Y.: Dover, 1962.

Newton, Isaac. *Mathematical Principles of Natural Philosophy and System of the World*. Edited by F. Cajori. Berkeley, Cal.: University of California Press, 1934.

Popper, Karl R. *The Logic of Scientific Discovery*. N.Y.: Harper & Row, 1959.

————. *The Poverty of Historicism*. N.Y.: Harper & Row, 1957.

Scott, Joseph Frederick. *The Scientific Work of René Descartes*. London: Taylor & Francis, 1952.

Science at the Cross Roads. Papers presented to the International Congress of the History of Science and Technology, London, 1931. London: Kniga (England), 1931.

ARTICLES

Croce, Benedetto. "History and Chronicle." In *Theories of History*, edited by Patrick Gardiner, pp. 226–33. Glencoe, Ill.: Free Press, 1954.

Einstein, Albert. "On the Method of Theoretical Physics," The Herbert Spencer lecture delivered at Oxford, June 10, 1933. In *Ideas and Opinions*, pp. 270–76. N.Y.: Crown, 1954.

Feyerabend, Paul. "Comments on Hanson's 'Is There a Logic of Scientific Discovery?' " In *Current Issues in the Philosophy of Science*, edited by Herbert Feigl and Grover Maxwell, pp. 35–39. N.Y.: Holt, Rinehart & Winston, 1961.

————. "Explanation, Reduction, and Empiricism." In *Scientific Explanation, Space, and Time,* edited by Herbert Feigl and Grover Maxwell. *Minnesota Studies in the Philosophy of Science,* vol. 3, pp. 28–97. Minneapolis: University of Minnesota Press, 1962.

————. "Reply to Criticism." In *Boston Studies in the Philosophy of Science,* edited by Robert S. Cohen, and Marx W. Wartofsky, vol. 2, pp. 223–61. N.Y.: Humanities Press, 1965.

————. "Review of Hanson," *Philosophical Review,* 73:264–66, 1964.

Gillispie, Charles C. "Review of Agassi's 'Towards an Historiography of Science'," *Isis,* 55:97–99, 1964.

Guerlac, Henry. "The Continental Reputation of Stephen Hales," *Archives Internationales des Sciences,* 4:393–404, 1951.

————. "A Lost Memoir of Lavoisier," *Isis,* 50:125–29, 1959.

Hahn, Roger. "Reflections on the History of Science," *Journal of the History of Philosophy,* 3:235–42, 1965.

Hall, A. Rupert. "Merton Revisited, or Science and Society in the Seventeenth Century," *History of Science,* 2:1–15, 1963.

————. "Merton Revisited, or Science and Society in the Seventeenth Century." In *The Rise of Modern Science: External or Internal Factors?,* edited by George Basalla, pp. 89–96. Lexington, Mass.: Heath, 1968.

Hempel, Carl G. "Reasons and Covering Laws in Historical Explanation." In *Philosophy and History,* edited by Sidney Hook, pp. 143–63. N.Y.: New York University Press, 1963.

Hessen, Boris M. "The Social and Economic Roots of Newton's Principia." In *Science at the Cross Roads.* Papers presented to the International Congress of the History of Science and Technology, London, 1931, by the representatives of the Soviet Union. London: Kniga (England), 1931.

Koyré, Alexandre. "Galileo and Plato," *Journal of the History of Ideas,* 4:400–28, 1943.

————. "Galileo and Plato," *Metaphysics and Measurement,* pp. 16–43. Cambridge, Mass.: Harvard University Press, 1968.

————. "Galileo and the Scientific Revolution in the Seventeenth Century," *Metaphysics and Measurement,* pp. 1–15. Cambridge, Mass.: Harvard University Press, 1968.

————. "Galileo and the Scientific Revolution in the Seventeenth Century," *Philosophical Review,* 52:333–48, 1943.

————. "Newton and Descartes," *Newtonian Studies,* pp. 53–114. London: Chapman & Hall, 1965.

————. "The Significance of the Newtonian Synthesis," Lecture, University of Chicago, November 3, 1948; reprinted in *Newtonian Studies,* pp. 3–24. London: Chapman & Hall, 1965.

Kuhn, Thomas S. "Review of Agassi's 'Towards an Historiography of Science'," *British Journal for the Philosophy of Science,* 17:256–58, November 1966.

―――. "Science: The History of Science." *International Encyclopedia of the Social Sciences*, 14:74–83.

Merton, Robert K. "Science, Technology, and Society in Seventeenth Century England," *Osiris*, 4:361–632, 1938.

Popper, Karl R. "The Aim of Science," *Ratio*, 1:24–35, December 1967.

Scriven, Michael. "Causes, Conditions, and Connections in History." In *Philosophical Analysis and History*, edited by William Dray, pp. 238–64. N.Y.: Harper & Row, 1966.

―――. "Explanation and Prediction in Evolutionary Theory," *Science*, 130: 477–82, 1959.

―――. "Explanations, Predictions, and Laws." In *Scientific Explanation, Space, and Time*, edited by Herbert Feigl and Grover Maxwell. *Minnesota Studies in the Philosophy of Science*, vol. 3, pp. 170–230. Minneapolis: University of Minnesota Press, 1962.

―――. "The Key Property of Physical Laws—Inaccuracy." In *Current Issues in the Philosophy of Science*, edited by Herbert Feigl and Grover Maxwell, pp. 91–104. N.Y.: Holt, Rinehart & Winston, 1961.

―――. "Science: The Philosophy of Science," *International Encyclopedia of the Social Sciences*, 14:83–92.

―――. "Truisms as the Grounds for Historical Explanations." In *Theories of History*, edited by Patrick Gardiner, pp. 443–75. Glencoe, Ill.: Free Press, 1959.

index

Index

Maurice A. Finocchiaro received his B.S. degree (1964) from the Massachusetts Institute of Technology, and his Ph.D. degree (1969) from the University of California, Berkeley. He is an assistant professor of philosophy at the University of Nevada, Las Vegas.

The manuscript was edited by Jean Spang. The book was designed by Don Ross. The typeface for the text is Linotype Caledonia designed by W. A. Dwiggins in 1938; and the display face is Eurostile designed by A. Novarese in 1962.

The text is printed on Oxford's Booklet Text paper and the book is bound in Columbia Mills' Fictionette cloth over binders' boards. Manufactured in the United States of America.